Design and Maintenance of a Network for Collecting High-Resolution Suspended-Sediment Data at Remote Locations on Rivers, with Examples from the Colorado River

By Ronald E. Griffiths, David J. Topping, Timothy Andrews, Glenn E. Bennett, Thomas A. Sabol, and Theodore S. Melis

Chapter 2 of
Section C, Instruments for Measurement of Suspended Sediment
Book 8, Instrumentation

Techniques and Methods 8–C2

U.S. Department of the Interior
U.S. Geological Survey

U.S. Department of the Interior
KEN SALAZAR, Secretary

U.S. Geological Survey
Marcia K. McNutt, Director

U.S. Geological Survey, Reston, Virginia 2012

This report and any updates to it are available online at:
http://pubs.usgs.gov/tm/tm8c2/

For more information on the USGS—the Federal source for science about the Earth, its natural and living resources, natural hazards, and the environment, visit http://www.usgs.gov or call 1-888-ASK-USGS (1-888-275-8747).

For an overview of USGS information products, including maps, imagery, and publications, visit http://www.usgs.gov/pubprod

To order this and other USGS information products, visit http://store.usgs.gov

Suggested citation:
Griffiths, R.E., Topping, D.J., Andrews, Timothy, Bennett, G.E., Sabol, T.A., and Melis, T.S., 2012, Design and maintenance of a network for collecting high-resolution suspended-sediment data at remote locations on rivers, with examples from the Colorado River: U.S. Geological Survey Techniques and Methods, book 8, chapter C2, 44 p.

Contents

Figures

Tables

Conversion Factors and Abbreviations

Multiply	By	To obtain
centimeter (cm)	0.3937	inch (in.)
millimeter (mm)	0.03937	inch (in.)
micron (μ)	0.00003937	inch (in.)
meter (m)	3.281	foot (ft)
kilometer (km)	0.6214	mile (mi)
liter (L)	0.2642	gallon (gal)
cubic meter per second (m^3/s)	35.31	cubic foot per second (ft^3/s)

Concentration of sediment in water is given in milligrams per liter (mg/L).
Horizontal coordinate information is referenced to the North American Datum of 1983 (NAD 83).

Additional Abbreviations

ABS	acrylonitrile butadiene styrene
ADCP	acoustic-Doppler current profiler
ADP	acoustic-Doppler profiler
BIOS	basic input/output system
EDI	equal-discharge-increment
EWI	equal-width-increment
FTP	file transfer protocol
GCMRC	Grand Canyon Monitoring and Research Center
GCNP	Grand Canyon National Park
kHz	kilohertz
LISST	Laser In-Situ-Scattering and Transmissometry
MHz	megahertz
PC	personal computer
RS–232	recommended standard 232
RTU	remote terminal unit
RUG	RUGID Computer RTU
SDI–12	serial digital interface at 1,200 baud rate
USB	Universal Serial Bus
USGS	U.S. Geological Survey

Design and Maintenance of a Network for Collecting High-Resolution Suspended-Sediment Data at Remote Locations on Rivers, with Examples from the Colorado River

By Ronald E. Griffiths, David J. Topping, Timothy Andrews, Glenn E. Bennett, Thomas A. Sabol, and Theodore S. Melis

Abstract

Management of sand and finer sediment in fluvial settings has become increasingly important for reasons ranging from endangered-species habitat to transport of sediment-associated contaminants. In all rivers, some fraction of the suspended load is transported as washload, and some as suspended bed material. Typically, the washload is composed of silt-and-clay-size sediment, and the suspended bed material is composed of sand-size sediment. In most rivers, as a result of changes in the upstream supply of silt and clay, large, systematic changes in the concentration of the washload occur over time, independent of changes in water discharge. Recent work has shown that large, systematic, discharge-independent changes in the concentration of the suspended bed material are also present in many rivers. In bedrock canyon rivers, such as the Colorado River in Grand Canyon National Park, changes in the upstream tributary supply of sand may cause large changes in the grain-size distribution of the bed sand, resulting in changes in both the concentration and grain-size distribution of the sand in suspension. Large discharge-independent changes in suspended-sand concentration coupled to discharge-independent changes in the grain-size distribution of the suspended sand are not unique to bedrock canyon rivers, but also occur in large alluvial rivers, such as the Mississippi River. These systematic changes in either suspended-silt-and-clay concentration or suspended-sand concentration may not be detectable by using conventional equal-discharge- or equal-width-increment measurements, which may be too infrequently collected relative to the time scale over which these changes in the sediment load are occurring. Furthermore, because large discharge-independent changes in both suspended-silt-and-clay and suspended-sand concentration are possible in many rivers, methods using water discharge as a proxy for suspended-sediment concentration (such as sediment rating curves) may not produce sufficiently accurate estimates of sediment loads. Finally, conventional suspended-sediment measurements are both labor and cost intensive and may not be possible at the resolution required to resolve discharge-independent changes in suspended-sediment concentration, especially in more remote locations. For these reasons, the U.S. Geological Survey has pursued the use of surrogate technologies (such as acoustic and laser diffraction) for providing higher-resolution measurements of suspended-sediment concentration and grain size than are possible by using conventional suspended-sediment measurements alone. These factors prompted the U.S. Geological Survey's Grand Canyon Monitoring and Research Center to design and construct a network to automatically measure suspended-sediment transport at 15-minute intervals by using acoustic and laser-diffraction surrogate technologies at remote locations along the Colorado River within Marble and Grand Canyons in Grand Canyon National Park. Because of the remoteness of the Colorado River in this reach, this network also included the design of a broadband satellite-telemetry system to communicate with the instruments deployed at each station in this network. Although the sediment-transport monitoring network described in this report was developed for the Colorado River in Grand Canyon National Park, the design of this network can easily be adapted for use on other rivers, no matter how remote.

In the Colorado River case-study example described in this report, suspended-sediment concentration and grain size are measured at five remote stations. At each of these stations, surrogate measurements of suspended-sediment concentration and grain size are made at 15-minute intervals using an array of different single-frequency acoustic-Doppler side-looking profilers. Laser-diffraction instruments are also used at two of these stations to measure both suspended-sediment concentrations and grain-size distributions. Cross-section calibrations of these instruments have been constructed and verified by using either equal-discharge-increment (EDI) or equal-width-increment (EWI) measurements of the velocity-weighted suspended-sediment concentration and grain-size distribution. The suspended-silt-and-clay concentration parts of these calibration relations have also included information from EDI- or EWI-calibrated samples collected by automatic pump samplers. Three of the monitoring stations are equipped with two-way satellite broadband telemetry systems that operate once a day to remotely monitor and program the instruments and download data. Data from these stations are typically downloaded twice per month; data from stations without satellite-telemetry systems are downloaded during site visits, which occur every 2 months or semiannually, depending on the remoteness of the site. Upon downloading and processing, suspended-silt-and-clay concentration, suspended-sand concentration, and suspended-sand median grain size are posted on the World Wide Web. Satellite telemetry in combination with the high-resolution sediment surrogate measurements can generate near-real-time suspended-sediment-concentration and grain-size data (limited only by the time required to download the instruments and process the data). The approach for measuring suspended-sediment concentration and grain size using this monitoring network is more practical, and can be done at a much lower cost and with higher temporal resolution, than any other method.

Introduction

The need to accurately monitor sediment transport in rivers has grown in importance in response to an array of environmental concerns, ranging from the degradation of endangered-species habitat to the transport and sequestration of sediment-bound contaminants (for example, U.S. Environmental Protection Agency, 2004; Owens and others, 2005; Larsen and others, 2010). In some rivers, upstream land use and water development have affected habitat by increasing the sediment supply relative to the sediment-transport capacity, whereas in other rivers, they have affected habitat by increasing the transport capacity relative to the supply (Wood and Armitage, 1997; Topping, Rubin, and Vierra, 2000; Topping, Rubin, and others, 2000; Grant and others, 2003; Schmidt and Wilcock, 2008). The first of these two cases is one in which the sediment budget is in surplus, and the second is one in which the sediment budget is in deficit. For example, in the Pacific Northwest and in parts of Europe, upstream logging and other land use practices have increased the loads and bed storage of sand and finer sediment in rivers, adversely affecting salmonid spawning habitat (Madej and Ozaki, 1996; Lisle and Napolitano, 1998; Greig and others, 2005). Conversely, in the Colorado River in Grand Canyon National Park (GCNP), construction and operation of Glen Canyon Dam has cutoff most of the sand supply, increased baseflow, and virtually eliminated floods (Topping, Rubin, and Vierra, 2000, Topping, Rubin, and others, 2000; Topping and others, 2003). These changes have resulted in substantial erosion of sandbars deemed important because they were an integral part of the natural riverscape upon creation of GCNP and provide riparian habitat, endangered-native-fish habitat, protection of archaeological sites, and recreation (Dolan and others, 1974; U.S. Department of the Interior, 1995; Rubin and others, 2002; Schmidt and others, 2004; Wright and others, 2005, 2008). In addition, upcoming major dam-removal projects will result in

the shift in the condition of reaches downstream from these dams from sediment deficit to sediment surplus (Randle and Bountry, 2010). Although the desired long-term benefit of these dam-removal projects is the restoration of salmonid habitat upstream from these dams, these projects may adversely affect the water quality in downstream reaches over shorter timescales. Finally, transport of sediment-bound contaminants is of concern in almost all rivers (U.S. Environmental Protection Agency, 2004).

Some fraction of the suspended-sediment load in all rivers is transported as washload, in which concentrations vary mostly as a function of changes in the upstream supply. The rest of the suspended-sediment load is transported as suspended bed material, in which concentrations vary largely as a function of changes in the discharge of water (Guy, 1964, 1970; Nordin and Beverage, 1965; Porterfield, 1972; Topping, Rubin, and Vierra, 2000; Topping, Rubin, and others, 2000; Rubin and Topping, 2001, 2008). In most rivers, the washload is composed of silt-and-clay-size sediment and the suspended bed material is composed of sand-size sediment. In some rivers, such as the dam-regulated Colorado River in GCNP, where the upstream supply of sand is limited and discharges are relatively high, changes in the upstream supply of sand (manifested mainly through changes in the grain-size distribution of the bed sand) can be as important as changes in the discharge of water in regulating suspended-sand transport (Rubin and Topping, 2001, 2008). During floods on upstream tributaries, the upstream supply of sand, silt, and clay is temporarily enriched in the dam-regulated Colorado River, resulting in large increases in suspended-sand and suspended-silt-and-clay concentrations (Topping, Rubin, and others, 2000); during these periods of sediment-supply enrichment, suspended-sediment concentrations can briefly approach the maximum values that occurred naturally before construction of Glen Canyon Dam. Some signature of changes in bed-sand grain size exerting a measurable control on suspended-sand concentration have even been detected in classical alluvial rivers that have a seemingly unlimited upstream supply of sand, such as the Mississippi River at Saint Louis, Missouri (Rubin and Topping, 2001, 2008). A factor of ~10 variation in suspended-sand concentration is possible at any given water discharge in the Mississippi River at Saint Louis, and a factor of ~1,000 variation in suspended-sand concentration is possible at any given water discharge in the Colorado River in Grand Canyon (fig. 1). In both of these rivers, this discharge-independent variation in suspended-sand concentration is coupled to systematic variation in the grain-size distribution of the suspended sand. In addition, because the suspended-silt-and-clay concentration is largely determined by changes in the upstream supply of silt and clay in both of these rivers, a factor of ~10 variation in suspended-silt-and-clay concentration is possible at any given water discharge in the Mississippi River at Saint Louis, and up to a factor of ~100,000 variation in suspended-silt-and-clay concentration is possible at any given water discharge in the Colorado River in Grand Canyon. The variation in both suspended-sand concentration and suspended-silt-and-clay concentration at any given discharge in the Mississippi and Colorado Rivers (fig. 1) is much larger than the errors associated with the suspended-sediment measurements (Topping and others, 2011). Therefore, this variation is real and, as shown by Topping, Rubin, and Vierra (2000); Topping, Rubin, and others (2000); and Rubin and Topping (2001), is systematic.

Because discharge-independent variation in either suspended-sand or suspended-silt-and-clay concentration is evident in most rivers, and when present is systematic and

can be more than a factor of 10 in magnitude, the discharge of water is a relatively poor proxy to use in estimating sand loads in most rivers and silt and clay loads in all rivers (as is commonly done using sediment rating curves). Therefore, the best approach to estimating sediment loads is to make discharge-independent, direct measurements of the concentration of suspended sediment at a temporal resolution sufficient to capture the systematic variation. In a large alluvial river like the Mississippi River, this required temporal resolution may be days

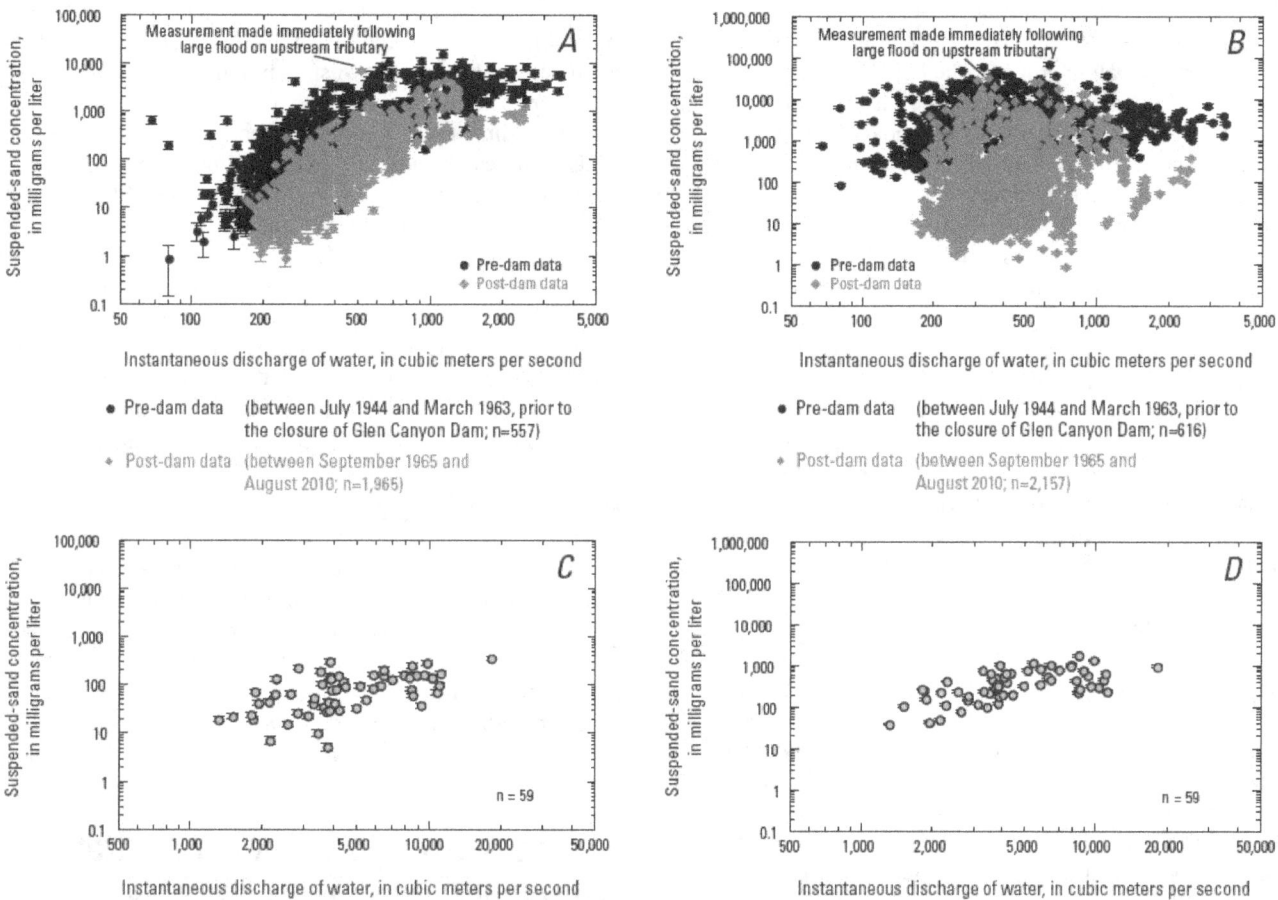

Figure 1. Plots depicting ranges in discharge-independent suspended-sediment concentration during the periods of grain-size analyzed sediment records in the Colorado and Mississippi Rivers. Variation in measured suspended-sand and suspended-silt-and-clay concentration at any given discharge is much larger than the errors associated with these measurements (Topping and others, 2011). A, Velocity-weighted suspended-sand concentration, and B, suspended-silt and clay concentration plotted as functions of water discharge at the U.S. Geological Survey streamflow-gaging station Colorado River near Grand Canyon, Arizona (09402500). Data were collected using isokinetic depth-integrating samplers deployed using either the equal-width- or equal-discharge-increment methods; see Edwards and Glysson, 1999. C, Velocity-weighted suspended-sand concentration, and D, suspended-silt and clay concentration plotted as functions of water discharge at U.S. Geological Survey streamflow-gaging station Mississippi River at Saint Louis, Missouri (07010000). Errors bars indicate the 95-percent-confidence-interval field error estimated using methods described in Topping and others (2011). Pre-dam Colorado River data were collected between July 1944 and March 1963; post-dam Colorado River data were collected between September 1965 and August 2010 (U.S. Geological Survey, 1947, 1972, 1976, 2011a; Paulsen, 1949, 1950, 1952a, 1952b, 1953; Love, 1954, 1955, 1957, 1958, 1959a, 1959b, 1960, 1961, 1964a, 1964b, 1966; Garrett and others, 1993; Topping and others, 1999, 2011; Topping, Rubin, and Melis, 2007; Topping, Rubin, and others, 2000; post-1998 data were collected as part of the case study described in this report). Mississippi River data were collected between December 1959 and March 1985 (Scott and Stephens, 1966; U.S. Geological Survey, 2011b).

to months, but in a bedrock canyon river like the Colorado River, this required temporal resolution is less than 1 hour (Topping, Rubin, and Vierra, 2000; Topping, Rubin, and others, 2000; Topping, Wright, and others, 2006, 2007). In either case, it is likely that the required resolution for suspended-sediment data collection is high enough to be labor and cost prohibitive when only conventional sampling methods, such as the equal-discharge-increment (EDI) or equal-width-increment (EWI) methods described in Edwards and Glysson (1999), are used. To resolve this sampling problem, the U.S. Geological Survey (USGS) has pursued the use of surrogate technologies (Gray and others, 2003; Gray and Gartner, 2009).

The Colorado River in GCNP is currently the focus of a major river restoration program, the U.S. Department of the Interior's Glen Canyon Dam Adaptive Management Program (Campbell and others, 2010). A major goal of this program is the restoration of eddy sandbars and other deposits composed of sand and finer sediment (that is, silt and clay) in and along the Colorado River in Marble and Grand Canyons. Maintenance of these sandbars and other deposits by resource managers through manipulation of dam releases following tributary inputs of new sediment requires detailed knowledge of the status of sediment budgets within discrete reaches of the Colorado River in Marble and Grand Canyons (Rubin and others, 2002; Topping, Rubin, and others, 2006; Melis and others, 2007; Topping and others, 2010). Because the upstream supply of sand and finer sediment is a strong regulator of sediment-transport rates in the Colorado River, and this supply of sediment can change substantially over timescales of hours, a high-resolution method for measuring sediment transport was deemed necessary to provide data sufficient to incorporate in management decisions.

To monitor sediment transport in the Marble and Grand Canyons reaches of the Colorado River, the USGS Grand Canyon Monitoring and Research Center designed and evaluated a network of monitoring stations that automatically measure suspended-sediment concentration and grain-size at high temporal resolutions using acoustic and laser-diffraction surrogate technologies, augmented by automatic pump samplers (Melis and others, 2003; Topping and others, 2004; Topping, Wright, and others, 2006, 2007). This monitoring network is composed of five sediment-transport monitoring stations operated on the Colorado River in Marble and Grand Canyons (fig. 2), some at extremely remote locations in GCNP that are accessible only by hiking long distances over rough terrain or by taking multiple-day boat trips. As part of the development of this monitoring network, broadband satellite-telemetry systems were designed to allow frequent communication with the acoustic, laser-diffraction, and pump-sampler instruments. A requirement of the design of the satellite-telemetry system was that the system allowed use of the software supplied by the vendors of these various instruments so that the instruments could be easily reprogrammed or the data downloaded from the office. Another component of this monitoring network was the design of a programmable

"trigger" to automatically collect pump samples on the basis of threshold values in the laser-diffraction data or any water-quality parameter (for example, turbidity, specific conductance, water temperature). Data retrieved from the monitoring stations with the satellite-telemetry systems are used to provide near real-time information on sediment transport in the Colorado River in GCNP to the resource managers in the Glen Canyon Dam Adaptive Management Program. The five monitoring stations, coupled with tributary monitoring sites described elsewhere (for example, Topping and others, 2010; Griffiths and others, 2010), compose a monitoring network to provide information to resource managers on the quantity and grain-size distribution of sand and finer sediment as the sediment is transported downstream in the Colorado River in Marble and Grand Canyons.

The sediment-transport monitoring network described in this report can be adapted for use on any river, no matter how remote. Very few major rivers in the United States are, in fact, as remote as is most of the Colorado River in GCNP. For example, some of the stations in this network can be visited only twice per year because they are so difficult to reach. In addition, the locations of the monitoring stations on the Colorado River are subject to an extremely large range in weather conditions and temperatures. Thus, the Colorado River in GCNP was the ideal location to develop and test the sediment-transport monitoring network described in this report.

Purpose and Scope

This report describes and documents the design and maintenance of a network of sediment-transport monitoring stations that was developed for the Colorado River in GCNP. This report can provide guidance for the design and maintenance of sediment-transport monitoring networks in remote settings on other rivers in addition to guidance for the maintenance of this specific sediment-transport monitoring network.

Sediment-Transport Monitoring Stations in the Colorado River Case Study

Five sediment-transport monitoring stations are in operation on the Colorado River in Marble and Grand Canyons (fig. 2 and table 1). Selection of monitoring station locations was based on the need to resolve longitudinal differences in sediment storage, to bracket major tributaries, and to reoccupy former USGS streamflow-gaging stations with historical sediment-transport data. These sediment-transport monitoring stations use multifrequency acoustics and laser diffraction to collect surrogate measurements of suspended-sediment concentration and grain size at 15-minute intervals (Melis and others, 2003; Topping and others, 2004; Topping, Wright, and others, 2006, 2007). Arrays of multifrequency side-looking acoustic-Doppler profilers (ADPs) and laser-diffraction

Table 1. Sediment-transport monitoring stations in the Colorado River case study.

[USGS, U.S. Geological Survey. River miles are used as place names and to refer to locations along the Colorado River in the study area. River miles increase downstream beginning at river mile 0, which is at the USGS streamflow-gaging station Colorado River at Lees Ferry, Arizona]

Formal station name	Station name used in this report
River-mile 30 sediment station	30-mile station
Former USGS streamflow-gaging station Colorado River above Little Colorado River near Desert View, Ariz. (09383100)	61-mile station
USGS streamflow-gaging station Colorado River near Grand Canyon, Ariz. (09402500)	87-mile station
Former USGS streamflow-gaging station Colorado River above National Canyon near Supai, Ariz. (09404120)	166-mile station
USGS streamflow-gaging station Colorado River above Diamond Creek near Peach Springs, Ariz. (09404200)	225-mile station

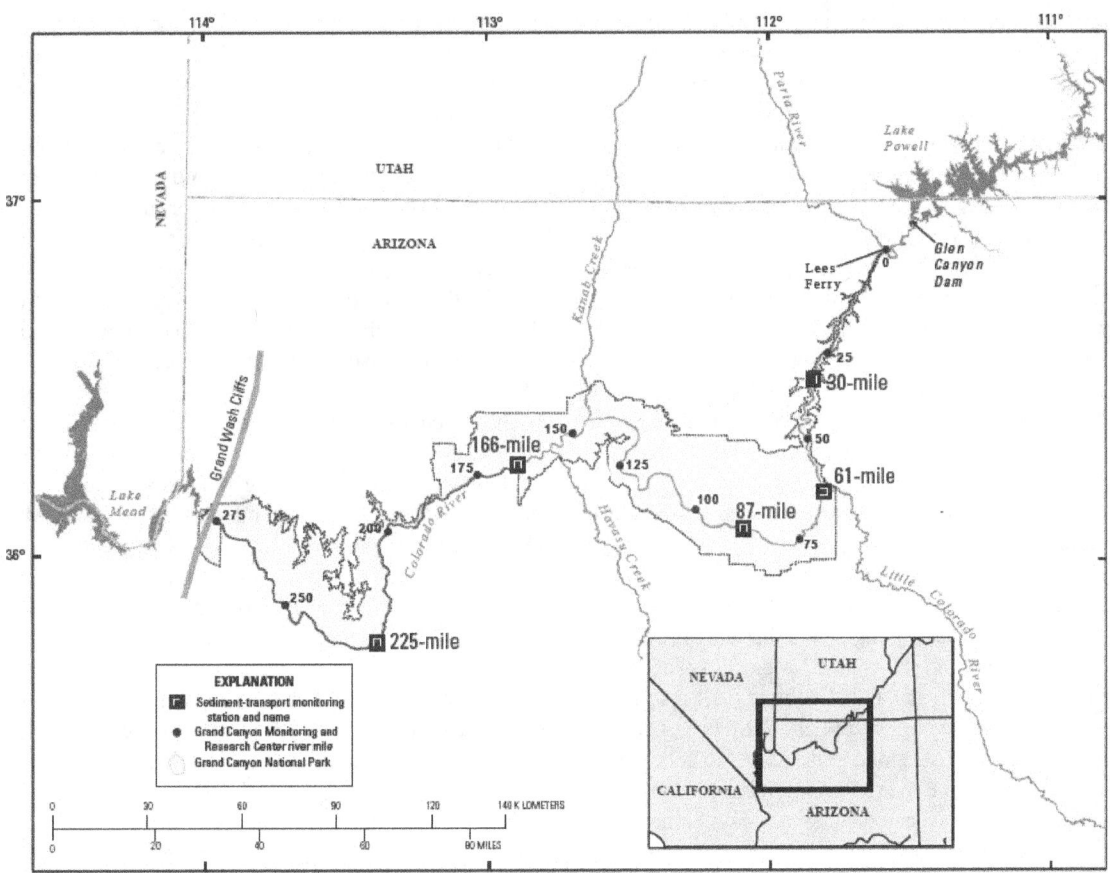

Figure 2. Locations of the five sediment-transport monitoring stations on the Colorado River in Marble and Grand Canyons. (See table 1 for station descriptions.)

instruments at these stations have been calibrated to measure the velocity-weighted suspended-sediment concentration and grain-size distribution in the adjacent river cross section by using either EDI or EWI measurements made with depth-integrating suspended-sediment samplers. In most cases, the silt-and-clay-concentration parts of these calibrations are augmented by using samples collected by ISCO automatic pump samplers. The result is a high-frequency record of suspended-silt-and-clay concentration, suspended-sand concentration, and either the suspended-sand median grain size (when ADP arrays are used) or the suspended-sand grain-size distribution (when laser diffraction is used) at each site. In addition to providing suspended-sediment measurements, the ADPs also measure river stage at 15-minute intervals. The ADP river-stage data are used in combination with episodic discharge measurements to compute water discharge at the three monitoring stations (30-mile, 61-mile, and 166-mile) not located at active USGS streamflow-gaging stations. By long-standing convention, river miles are used as place names and to refer to locations along the Colorado River in the study area. All river miles referred to in this report are USGS Grand Canyon Monitoring and Research Center river miles; river-mile 0 is at the USGS streamflow-gaging station Colorado River at Lees Ferry, Arizona (09380000).

Three of the five monitoring stations (30-mile, 61-mile, and 87-mile) are equipped with two-way satellite-telemetry systems that automatically operate once a day to provide remote monitoring and programming of the instruments, and to allow for data downloading. In addition to providing programming and downloading capabilities, the two-way satellite-telemetry systems allow vendor-provided software specific to the instrument, as opposed to a generic interface such as HyperTerminal, to be used when interfacing with instruments. Data from the three stations equipped with two-way satellite-telemetry systems are typically downloaded every other week, whereas data from the other two stations are retrieved during site visits (usually twice a year at 166-mile and once every two months at 225-mile). If needed, data from stations equipped with two-way satellite-telemetry systems can be downloaded on a daily basis to provide near real-time information on suspended-sediment concentrations and grain-size distributions at these locations.

The primary instruments used to monitor suspended-sediment concentration and grain size at the sediment-transport monitoring stations are side-looking ADPs manufactured by Nortek and OTT. ADPs used are 1- and 2-megahertz (MHz) EasyQ River Flow Monitors and 600-kilohertz (kHz) Aquadopp Profilers manufactured by Nortek and 2-MHz SLDs manufactured by OTT Messtechnik.[1] These instruments are rigidly mounted to either the riverbank or gaging-station structures below the minimum expected stage based on permitted dam operations (the stage associated with a water discharge

of 142 cubic meters per second (m3/s)). Methods used to relate multifrequency acoustic attenuation and backscatter to suspended-sediment concentration and grain size are described in Topping, Wright, and others (2006, 2007) and Wright, Topping, and Williams (2010). Data are collected and logged on the internal dataloggers in each ADP every 15 minutes.

The secondary instruments used to monitor suspended-sediment concentration and grain size are Laser In-Situ-Scattering and Transmissometry (LISST) laser-diffraction instruments manufactured by Sequoia Scientific. LISST–100 type C instruments are deployed at the 61-mile and 87-mile stations. These instruments provide a point measurement of the concentration of suspended-sediment in 32 logarithmically spaced size classes between 2.5 and 500 microns (Agrawal and Pottsmith, 1994, 2000).

The ADP and LISST–100 instruments have been calibrated to the velocity-weighted concentration of suspended-sediment (in various size classes) in the adjacent river cross section by using EDI and EWI measurements made at each station with depth-integrating suspended-sediment samplers (measurement methods described in Edwards and Glysson, 1999). Most historical, and all current (2011), instrument calibrations have been developed using US D–96–A1 or US D–96 samplers (Davis, 2001; Federal Interagency Sedimentation Project, 2003).

To aid in the development of these calibration relations, especially during periods of high suspended-silt-and-clay concentration when EDI or EWI data are typically sparse, ISCO 6712 automatic pump samplers at the sediment-transport monitoring stations are threshold triggered using either LISST–100 laser-transmission data or turbidity data collected by a YSI 6920 multiparameter water-quality instrument (sonde). Silt-and-clay-concentration parts of these ADP and LISST–100 calibration relations thus also include data from EDI- or EWI-calibrated samples collected by these automatic pump samplers. Sand-concentration parts of the ADP and LISST-100 calibration relations are developed on the basis of the more-accurate EDI and EWI measurements.

ISCO 6712 automatic pump samplers at four of the five sediment-transport monitoring stations are used to collect calibration verification suspended-sediment samples during periods of high suspended silt-and-clay concentration arising from active flooding on upstream tributaries and, if a second sampler is present, also at a fixed interval (usually 96 hours).[2] The ISCO 6712 automatic pump sampler uses a peristaltic pump to collect water samples though a 0.79-centimeter (cm) inside-diameter vinyl tubing. It deposits each sample in one of 24 one-liter (L) bottles in a carousel and records the time and date when the sample was collected. The pump sampler can theoretically pump to head heights of 8.5 meters (m) with intake tube lengths of as much as 30 m while maintaining velocities sufficiently high to adequately sample suspended

---

[1] OTT Messtechnik took over production of EasyQ ADPs from Nortek in 2006.

[2] A pump sampler is not typically deployed at the 166-mile station because the remoteness of this station precludes easy access to maintain the sampler.

Table 2. Equipment present at each sediment-transport monitoring station on the Colorado River in Marble and Grand Canyons

[ADPs, side-looking acoustic-Doppler profilers; LISSTs, Laser In-Situ-Scattering and Transmissometry instruments; ISCO, ISCO 6712 automatic pump sampler; RUG, RUGID Computer remote terminal unit; MHz, megahertz; kHz, kilohertz; SDI-12, serial data interface using 1,200 baud rate; NA, not applicable]

Station name	Frequency of ADPs on site	LISSTs on site	Number of ISCOs	ISCO trigger method	Number of YSI sondes	Satellite telemetry present?
30-mile	1 MHz, 2 MHz	None	2	YSI via RUG	2	Yes
61-mile	1 MHz, 2 MHz	LISST–100	2	YSI via RUG	2	Yes
87-mile	600 kHz, 1 MHz, 2 MHz	LISST–100	1	LISST–100	1	Yes
166-mile	1 MHz, 2 MHz	None	0	NA	0	No
225-mile	1 MHz, 2 MHz	None	2	YSI via SDI-12	2	No

sand. However, because of the greater amount of water in the intake tube relative to the 1-L sample volume, tube lengths much greater than 10 m result in substantial grain-size segregation in the intake tube, causing undersampling of sand-sized particles. At the one station where an intake tube substantially longer than 10 m is required (61-mile), a 0.64-cm inside-diameter intake tube is used to reduce the effect of this problem. The pump samplers connect to the satellite-telemetry system, if present, and operate on 12 volts.

The threshold triggered ISCO pump samplers are set to collect samples when either the turbidity measured by a YSI sonde exceeds a user-defined threshold or the laser transmission measured by a LISST–100 drops below a user-defined threshold. At three stations (30-mile, 61-mile, and 225-mile), a second sampler is present to also collect samples at a programmed interval, usually 96 hours. Collection of pump samples during episodes of high suspended-silt-and-clay concentration ensures that the suspended-sediment data record will be continuous if the measurement limits of the LISST–100s are exceeded. In addition, as mentioned above, these high-suspended-silt-and-clay-concentration pump samples help to both define and then subsequently verify the silt-and-clay concentration parts of the ADP instrument calibrations. This is necessary because the remoteness of all stations typically precludes more accurate EDI and EWI measurements from being collected during the relatively short periods of high suspended-silt-and-clay concentration in the Colorado River following tributary floods. Furthermore, collection of both the threshold-triggered and constant-interval pump samples helps ensure that the suspended-sediment data record will be continuous if the ADPs or LISST–100s fail.

Additional episodic EDI or EWI measurements made subsequent to the periods of instrument calibration are used to verify the calibration relations and compute out-of-sample errors.[3] As in the calibration process, additional episodic EDI-

or EWI-calibrated pump samples made subsequent to the periods of instrument calibration are used to aid in the verification of the high-silt-and-clay-concentration parts of the calibration relations and compute out-of-sample errors in suspended-silt-and-clay concentration. The processed ADP and LISST-100 suspended-sediment measurements are posted on the World Wide Web at http://www.gcmrc.gov.

Generalized Station Design

The generalized station design used at four of the five sediment-transport monitoring stations (30-mile, 61-mile, 87-mile, and 225-mile) in Marble and Grand Canyons consists of sediment-surrogate instruments and water-quality sonde(s) installed in the river, triggered automatic pump samplers (ISCOs), and 12-volt power supplies (table 2). The fifth sediment-transport monitoring station (166-mile) consists of sediment-surrogate instruments installed in the river and 12-volt power supplies. Instruments in the river—including the sediment-surrogate ADPs and laser-diffraction instruments, and water-quality instruments—are connected by communication/power cables to the power supply and, if present, to the satellite-telemetry system. At stations where local river geometry prevents ADPs from being deployed near the satellite-telemetry system (30-mile and 61-mile stations), Max-Stream (now named DIGI) radio modems are used to remotely connect the instruments to the satellite-telemetry system (fig. 3). At the three stations where ISCOs are threshold triggered using data collected by YSI sondes (30-mile, 61-mile, 225-mile), the ISCOs are either directly connected to a YSI sonde through an SDI-12 (serial data interface at 1,200 baud rate) cable or indirectly connected to a YSI sonde through a RUGID Computer remote terminal unit (RTU; herein referred to as a RUG). At the two stations where ISCOs can be threshold triggered using laser-transmission data (61-mile and 87-mile), ISCOs may also be directly connected to LISST-100s. All cable ends, power-supply components, and satellite-telemetry components are contained in weatherproof locked boxes

---

[3] Out-of-sample error is the error computed by using suspended-sediment data not included in the calibration dataset.

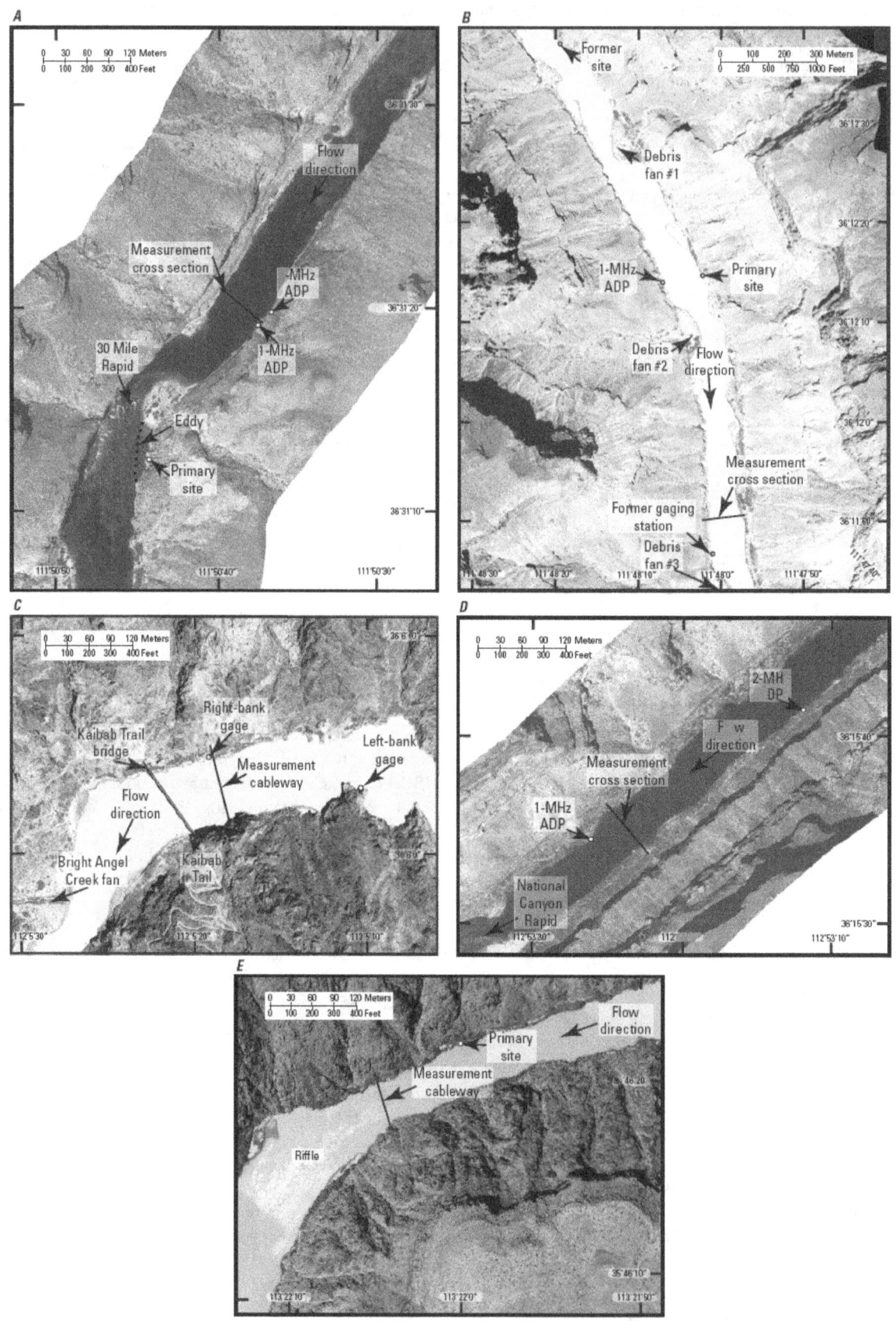

installed at elevations well above stages associated with normal dam operations and typically above the stages associated with controlled floods released from the dam to rebuild sandbars (for example, Hazel and others, 2010; Topping and others, 2010). Solar panels recharge 12-volt deep-cycle batteries that supply power at all sites. Figure 4 shows the generalized relation between the instruments at a monitoring station.

At all stations, the ADPs are mounted to the riverbank in a region where the flow is relatively uniform in the cross-stream dimension (that is, in a region outside of eddies where the flow is oriented downstream and varies smoothly away from the bank). Each ADP is mounted within 3 degrees of horizontal and below the stage associated with the lowest permitted release from Glen Canyon Dam. At all stations but the 87-mile station, each ADP is mounted on 3-inch-diameter (nominal) rigid acrylonitrile butadiene styrene (ABS) plastic pipe attached to bedrock or large boulders in the riverbank (fig. 5). The array of three single-frequency ADPs at the 87-mile station is mounted on a movable sled, which facilitates maintenance, and attached to a vertical aluminum H-beam bolted to the stilling well of the left-bank gage at the USGS streamflow-gaging station (fig. 6). The ADP mounts are typically L-shaped with a short (0.5 m) arm, to which the instruments are mounted. This short arm is glued to a longer (3–5 m) length of pipe attached to the riverbank. The ABS mounts are attached to either bedrock or boulders by using a combination of worm-gear band clamps, coated steel cable, pitons, angles, and in some cases expansion bolts (for a description of the ADP mounting procedures, see appendix 1.) To protect the communication and power cable from river-borne debris, the cable is strung through the ABS pipe mount. To prevent catastrophic loss of data from vandalism or rockfall, the different frequency ADPs at the three more-remote stations (30-mile, 61-mile, and 166-mile) are mounted at separate locations within the reaches.

Because ADPs are not greatly affected by biofouling, they typically require only minimal maintenance. Periodic (every two months or semiannually, depending on the station) wiping of the transducers to remove deposited sediment is usually all that is required to keep these instruments operating smoothly and collecting the highest-quality data.

At the 61-mile and 87-mile stations, LISST–100s are suspended below the water surface, and at least a meter above the bed of the river, from coated steel cables that are anchored to the bank on both ends. The LISST–100s are attached horizontally to the middle of the length of coated cable by using worm-gear band clamps. Unlike the ADPs, the LISST–100s are subject to biofouling and therefore must be cleaned periodically to ensure the collection of high-quality data for extended periods. Biofouling of the optics can occur at varying rates—depending on the season, water clarity, and amount of sunlight reaching the instrument—resulting in data that are unusable after periods of days to weeks. Unlike turbidity probes (that can be kept clean by using wipers), automatic wipers cannot be used to clean LISST optics because wipers would smear the glass lenses and alter the diffraction patterns. Therefore, the optics on LISST instruments must be kept clean by technicians visiting the station. To prevent sediment accumulation near the optics and reduce light-stimulated biofouling of the optics, an ABS light shield has been installed on the upper surface of the optics end of the LISST–100 (fig. 7). Communication/power cables connect the instrument to the satellite-telemetry system.

Figure 3. Orthorectified aerial photographs of the sediment-transport monitoring stations showing the locations of instruments and measurement cross sections. *A*, 30-mile station (USGS photograph taken in 2002). Primary site—current location of satellite-telemetry system, ISCO pump samplers (two), and YSI sondes (two), and former location of a 1-MHz ADP. 30 Mile Rapid is the hydraulic control for the pool containing the 1-MHz and 2-MHz ADPs and the measurement cross section. *B*, 61-mile station (USGS photograph taken in 2005). Primary site—current location of satellite-telemetry system, LISST–100, 2-MHz ADP, ISCO pump samplers (two), and YSI sondes (two). Former site—inferior 60-mile station where instruments were located prior to September 2004. Debris fan #1 is the hydraulic control for the former site. Debris fan #2 is the hydraulic control for the primary site. Debris fan #3 is the hydraulic control for the former location of USGS streamflow-gaging station Colorado River above Little Colorado River near Desert View, Arizona (09383100). Measurement cross section is at the former location of the gaging station cableway. *C*, 87-mile station (USGS photograph taken in 2005). Two gages (left-bank and right-bank) compose the USGS streamflow-gaging station Colorado River near Grand Canyon, Arizona (09402500). Left-bank gage—location of satellite telemetry system, LISST–100, 1-MHz ADP, 2-MHz ADP, 600-kHZ ADP, ISCO pump sampler, and YSI sonde. Bright Angel Creek fan is the hydraulic control for the gaging station. *D*, 166-mile station (USGS photograph taken in 2002). The 1-MHz ADP is located at the former location of USGS streamflow-gaging station Colorado River above National Canyon near Supai, Arizona (09404120). Measurement cross section is at the former location of the gaging station cableway. *E*, 225-mile station (USGS photograph taken in 2005). Primary site—current location of 1-MHz ADP, 2-MHz ADP, ISCO pump samplers (two), and YSI sondes (two); adjacent to USGS streamflow-gaging station Colorado River above Diamond Creek, near Peach Springs, Arizona (09404200). Riffle (indicated) is the hydraulic control for the gaging station during normal operations of Glen Canyon Dam; during periods of higher discharge, the hydraulic control moves downstream approximately 600 meters to Diamond Creek Rapid.

As stated previously, the threshold-triggered ISCO 6712 automatic pump samplers at each sediment-transport monitoring station are triggered to collect samples during episodes of high suspended-silt-and-clay concentration that arise following periods of flooding on upstream tributaries. Two surrogate parameters for silt and clay concentration that can be used to trigger the pump samplers, and are easily measured, are turbidity and LISST laser transmission. Even though turbidity data are typically noisier than LISST laser-transmission data, turbidity probes can be kept relatively clean by wipers, and thus, a turbidity threshold is a more desirable parameter for triggering pump samplers at remote stations where it is logistically more difficult to clean the LISST optics. At the 30-mile, 61-mile, and 225-mile stations, turbidity measured by YSI sondes is therefore used to trigger the pump samplers. At the 30-mile and 61-mile stations, one of the two pump samplers is triggered by turbidity measured by a YSI sonde connected to the pump sampler through a USGS Grand Canyon Monitoring and Research Center designed system that uses a Rugid Computer running RUG3-Rev8 software (herein referred to as a RUG Trigger). The

RUG Trigger[4] is programmed to activate the pump sampler when the turbidity reaches, and remains at or above, a threshold for a set number of measurements. This approach prevents sampling during anomalous turbidity spikes. At the 225-mile station, one of the two pump samplers is triggered by turbidity measured by a YSI sonde connected directly to the pump sampler with an SDI–12 cable.

Accurate measurements of laser transmission by use of a LISST–100 require that the optics be cleaned at least monthly. Therefore, LISST–100 laser transmission is typically used to trigger the collection of pump samples only at the less-remote 87-mile station, which can be accessed by a well-traveled trail. Although laser transmission can also be used to trigger the collection of pump samples at the 61-mile station, this is not typically done because the remoteness of the station prevents frequent cleaning of the LISST–100, resulting in false triggering of the pump sampler.

---

[4]Although a turbidity threshold is the parameter chosen to trigger an ISCO pump sampler, the RUG Trigger is designed to activate a pump sampler by using any water-quality parameter that can be measured with a YSI sonde.

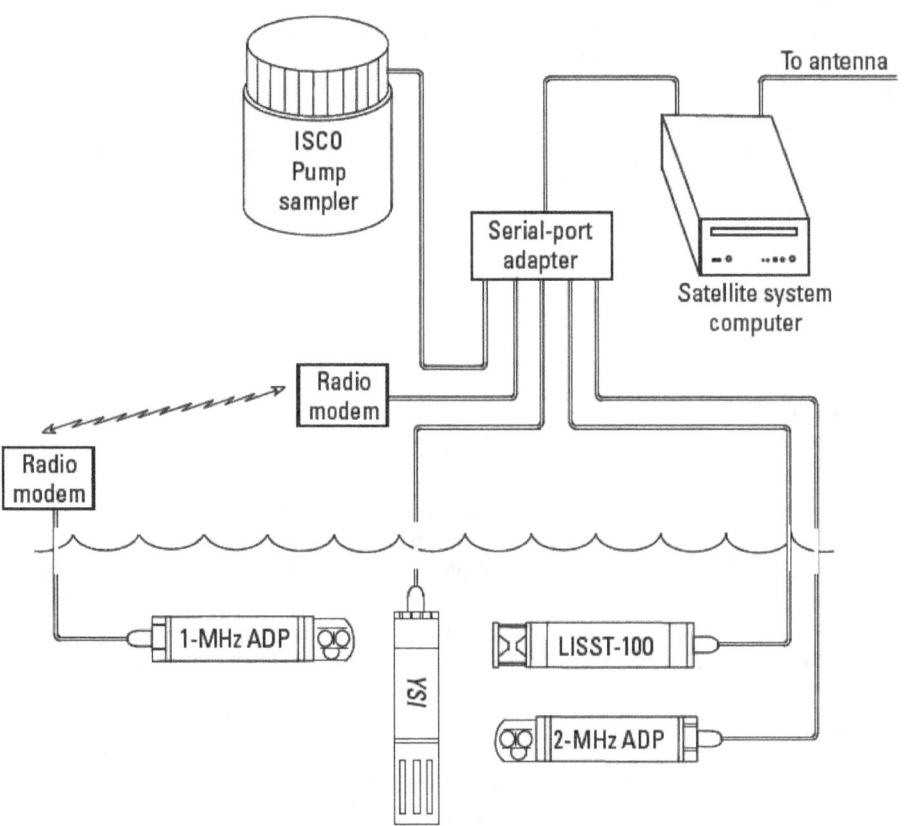

Figure 4. Simplified diagram of instrumentation at a sediment-transport monitoring station with satellite telemetry. The ISCO pump sampler may be triggered by either the YSI sonde or the LISST–100. The YSI sonde may be set up to trigger the ISCO pump sampler directly or through the RUG Trigger (not shown).

The solar power-supply systems at the sites are designed with some redundancy so that not all the instruments will be affected in the event of a power-system component failure. Therefore, each site has two to four independent charging circuits. At most sites, the individual instruments are connected to separate power supplies. In addition, satellite-telemetry systems are independently powered by 80-watt solar panels, charge controllers, and 12-volt deep-cycle batteries. Solar panels at some stations have been placed as much as 150 m away from the battery banks because of limited sunlight during winter in the depths of Marble and Grand Canyons. To minimize the effect of transmission loss, charge controllers have been installed on the battery side of these longer power cables.

Because the sediment-transport monitoring stations are within Grand Canyon National Park, the visibility of these stations needs to be minimized. At stations with preexisting USGS gaging-station infrastructure (87-mile and 225-mile), the visibility is of less concern; however, concealment of the equipment still provides protection against vandalism. Naturally occurring large boulders, artificial rock walls, camouflage paint, neutral-colored oiled-canvas tarps, and plant material conceal instruments and equipment installed at the three stations without preexisting gaging-station infrastructure (30-mile, 61-mile, and 166-mile; fig. 8). Solar panels, which are highly reflective, are positioned so that they are not visible from the river or are only minimally visible from downstream (most people on river trips do not look upstream). ADP mounts, constructed from ABS pipe, are covered with sand glued to the pipe with ABS glue; the sand coating breaks up the sharp profile of the pipe and mimics the wetting pattern of the surrounding rock (fig. 5).

Figure 5. Mount for the 2-MHz ADP at the 30-mile station. Photograph shows the camouflaged 3-inch-diameter (nominal) ABS plastic-pipe ADP mount and the barely visible radio modem antenna.

Figure 6. ADP mount at the 87-mile station. Photograph shows ADPs removed from the water for servicing.

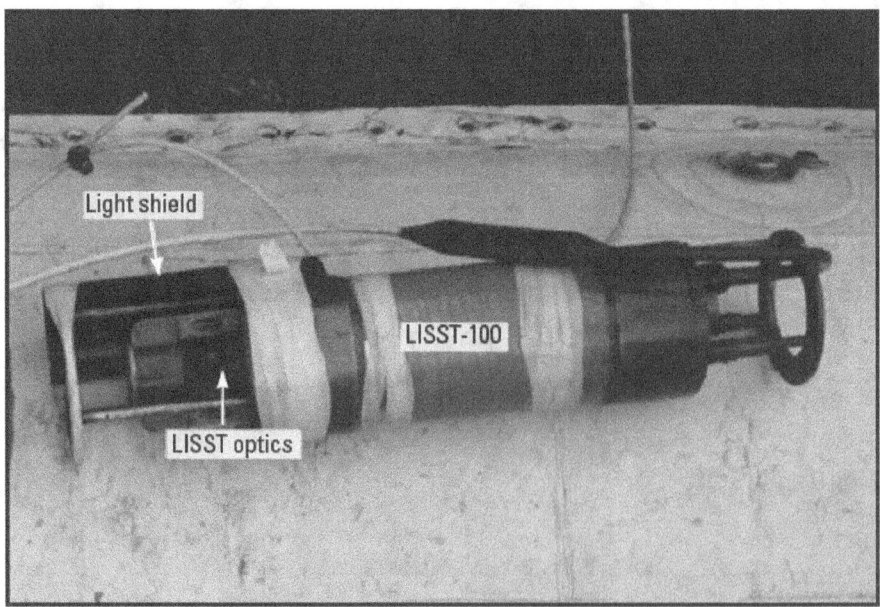

Figure 7. LISST–100 with ABS light shield installed. Photograph shows the instrument removed from the water for servicing.

Considerations for Monitoring-Station Placement

The locations of the sediment-transport monitoring stations are selected on the basis of the following criteria: distance between stations along the river and downstream from Glen Canyon Dam, location of important sediment-supplying tributaries, location of existing or former USGS streamflow-gaging stations, locations where historical sediment-transport data have been collected by the USGS, site accessibility, suitability of the site for mounting instruments in the river, year-round availability of sunlight for solar power, and local hydraulic conditions. The selected locations of the monitoring stations along the Colorado River in Marble and Grand Canyons allows for both monitoring of the sediment transported past each station and the calculation of changes in sediment storage in key reaches between stations.

For routine maintenance and repair of the instruments, access to the sediment-transport monitoring stations in this study is available by trail, boat, or gravel road, depending on the station. The most easily accessible of the five stations are the 87-mile and 225-mile stations; the most remote is the 166-mile station. Boat access to all monitoring stations is restricted by rapids on the Colorado River, making upstream travel impossible, and by National Park Service rules regulating the number of boat launches. Currently, stations are visited by boat twice a year. These river trips allow for major repairs, thorough maintenance, and collection of EDI or EWI measurements for instrument calibration and verification. More frequent collection of EDI measurements is only possible from the measurement cableways at the 87-mile and 225-mile stations. The 30-mile station is accessed by hiking an abandoned Bureau of Reclamation trail (approximately 12-km long with a 0.7-km elevation change) and by semiannual river trips. The 61-mile station is accessed by hiking the Walter Powell Route (an approximately 6.5-km long off-trail route with a 1-km elevation change that includes a class-4 rock climb and scramble from the canyon rim to the Little Colorado River, followed by a 2.5-km long cross-country hike to the site), by helicopter (the landing site is near the base of the Walter Powell Route on the Little Colorado River), and by semiannual river trips. The 87-mile station is accessed by hiking the South Kaibab Trail (approximately 12-km long with a 1.5-km elevation change) and by semiannual river trips. Although hiking distances are similar accessing the 30-mile and 87-mile stations, ease of access is greater at the 87-mile station because of the proximity of National Park Service facilities and housing. Access to the extremely remote 166-mile station is limited to semiannual river trips. The 225-mile station is accessed by gravel road followed by a 1.5-km long hike and by semiannual river trips.

In addition to general considerations for the placement of monitoring stations, substantial consideration must also be given to the placement of specific instruments. The hydraulic conditions present at each site affect the quality of the data collected and the maintenance required to keep the instruments functioning properly. Installation of ADPs requires both a suitable mounting location and favorable, relatively uniform hydraulic conditions. The considerations for selecting a mounting location for the ADPs include the following:

- The proximity and line of sight (for radio modem communication) from the proposed instrument site to the primary site (where the satellite telemetry and other instruments are deployed).

Figure 8. Instrumentation and site appearance at 30-mile sediment-transport station. *A,* Configuration of instruments. The rock wall behind the instruments is used to camouflage the station. *B,* Site appearance from river level. View is downstream. Site is concealed behind the rock wall.

- The availability of attachment points for the ADP mounts. Attachment points for the ABS pipe or aluminum mounts should be either solid rock or cement, prevent the instrument from changing orientation over time, and be strong and numerous enough to keep the instrument secure during periods of high discharge.

- The presence of objects such as boulders in the channel in front of the instrument that may block one or more of the acoustic beams. Because of beam spreading and side-lobe interference, objects not directly in front of the beams may partially block a beam.

- The presence, or potential presence, of intermittent sand bars that may episodically bury the instrument or block the instrument's acoustic beams.

The hydraulic-condition considerations for selecting a site for ADP instruments include the following:

- The horizontal acoustic beams are penetrating flow that is relatively uniform in the cross-stream dimension, in a region outside of eddies where the flow is oriented downstream, and that varies smoothly away from the bank.

- The suspended-sediment/water mixture to be sampled by the horizontal acoustic beams should be relatively well mixed without major step changes in both the suspended-sediment concentration and grain-size distribution along the beams.

- The acoustic beams should sample where there are few air bubbles in the water column, because air bubbles absorb the acoustic signal, resulting in poor data.

- The water velocity at the instrument head should be sufficient to prevent the buildup of large quantities of sediment. Sediment deposition on the transducers always results in degradation of the data. Because sediment buildup on the transducers decreases the intensity of the acoustic signal, a buildup of sediment may either require additional data processing (to compensate for the reduction in the acoustic signal) or, in some cases, render the data unusable.

LISST–100 instruments require two opposing anchors several meters apart, water reasonably representative of the average suspended-sediment conditions in the river cross section, and a location within the maximum communication cable length to the primary site (where the satellite-telemetry system and ISCO pump samplers are located). Unlike the ADPs, LISST instruments can be successfully deployed in eddies, although they should still be deployed in regions where few air bubbles are present in the water because LISST instruments cannot distinguish between air bubbles and sand-size particles. The ISCO pump samplers should be installed in areas where the water sampled is reasonably representative of the average suspended-sediment conditions in the river cross section, the samplers are above the level of expected high discharge, and the intake tube has a length of less than 10 m (to prevent substantial segregation of sand-sized particles in the intake tube that will lead to undersampling of suspended sand).

Two-Way Satellite-Telemetry System

Two-way satellite-telemetry systems, installed at three of the five sediment-transport monitoring stations, allow for the remote monitoring and programming of instruments, downloading of data from the instruments, and uploading of firmware and other updates to the instruments. These systems differ from the standard satellite-telemetry systems used at USGS streamflow-gaging stations in that the two-way satellite-telemetry systems do not use SDI–12 protocol and allow remote communication with any instrument by using the vendor-supplied software for that instrument. The stations chosen for installation of the satellite-telemetry systems are the upstream-most three stations (30-mile, 61-mile, and 87-mile). These stations were chosen because near real-time sediment conditions in the key reaches of the Colorado River bounded by these stations form the basis for management decisions on dam operations (data from the stations are typically downloaded twice a month but can be downloaded daily). The systems consist of an onsite Windows XP based computer (PC) running the Symantec remote control software "pcAnywhere," a second low-power-usage computer (RUG) that boots the system up each day at a user-defined time for a user-defined duration, a satellite modem, a satellite antenna, and a power supply (fig. 9). Instruments at each site are checked weekly, by using the satellite-telemetry system, to ensure that they are operating normally. (Instructions and information on checking and downloading sites remotely, as well as a detailed wiring connection diagram can be found in appendix 2.) At the sediment-transport monitoring stations where instruments are connected to the satellite-telemetry systems, data are downloaded as needed to the onsite PC and then transferred via satellite to the office, using file transfer protocol (FTP). Typically, the data from the ADPs and LISST-100s are processed to yield suspended-sediment concentrations and grain-size distributions, and posted to the World Wide Web every one to two months.

The onsite PC is programmed to automatically turn on once daily but can also be powered on directly at the site. Daily powering on of the system is done with a small low-power-usage computer that remains on all the time. This computer, a RUG (a separate RUG from the one used in the RUG Trigger to activate the ISCO pump samplers), monitors system temperatures and controls power to the PC and modem. The RUG, running Rugid Computer proprietary software, is programmed with setpoints indicating the time to turn the power on as well as conditions that warrant shutting the system off to prevent damage. The main system power is turned on by the RUG with the use of built-in relay switches. When these relay switches are closed, they in turn close other, higher-amperage relay switches (external to the RUG) that send power to the PC and modem. The system can also be powered on manually at the site using a RUG over-ride switch. Programming the setpoints into the RUG can be accomplished directly, through use of a built-in interface, or through the SatControl program on the PC; the SatControl program was created by USGS Grand Canyon Monitoring and Research Center personnel and contractors in Visual Basic using Visual Studio 2003. The RUG is connected to the PC with a serial cable that allows a remote operator to change the RUG setpoints and to download data stored on the RUG. Setpoints are changed remotely by using the PC's SatControl program. The SatControl program interfaces with the RUG and can use FTP to copy RUG data to and from the remote site. The PC, connected to the RUG and satellite modem, loads the SatControl program when it boots up.

The onsite PC is used to directly communicate with the instruments and to transfer instrument data from the site to the office. Instruments are checked by using either the instrument manufacturer's software or the Windows XP HyperTerminal program. Instruments, or the radio modems that connect to the instruments, are connected to a Quatech Universal Serial Bus (USB) multiport serial adapter with Recommended Standard 232 (RS–232) cables. The serial adapter allows multiple instruments, all of which use RS–232, to connect to the PC simultaneously through one USB port on the PC.

The SatControl program is used to keep the remote system power on, monitor the site temperatures, set the time the satellite-telemetry system turns on, shut down the system when battery voltages are too low, shut down the system when the temperature is too high, and maintain other controls. Each day, when the satellite-telemetry system turns on, the SatControl program is loaded on the PC and setpoints are sent to the RUG to set the next time the satellite-telemetry system will power on. Users can temporarily (for one startup cycle) or permanently change the setpoints in the SatControl program. (For a more detailed description of SatControl, see appendix 2.)

At the 87-mile station, lightning strikes and resultant electrical surges have caused problems with the instruments and satellite-telemetry system. This site may be susceptible to lightning strikes because of the large metal stilling well at the gaging station, the cables running from solar panels and a satellite dish installed on a hill above the site, and because of the presence of the historic Trans-Canyon Telephone line. To mitigate future damage, a lightning-surge protection system was installed at this site. The protection system consists of grounding, in-line surge protection, and optical isolators. A full description of the lightning-surge protection system and a wiring-connection diagram are given in appendix 2.

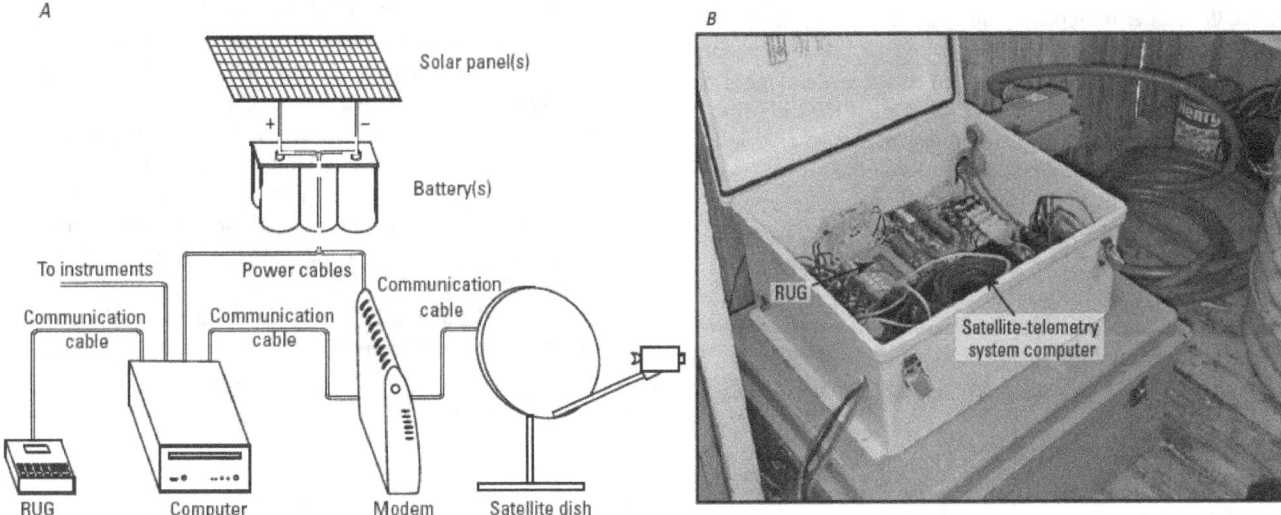

Figure 9. Two-way satellite-telemetry system. *A,* Simplified diagram. The 80-watt solar panel keeps the 12-volt battery(s) charged, which, in turn, supply power to the on-site PC and modem. Because of the limited power available on site and the high air temperatures during the summer months, the computer system is powered off during most of the day. The "wake-up" computer (Rugid RUG3) turns on the power to the computer and modem at a preset time; the computer system can also be powered on directly at the site. When the computer system is powered on, the computer can be controlled remotely over the Internet. Data can be downloaded, the instruments can be programmed and relaunched, and data files transferred by file transfer protocol (ftp) via satellite back to the controlling computer in the office. *B,* The 87-mile station satellite-telemetry system, including the RUG ("wake-up" computer), the on-site PC, and various relays used to switch on power to the system.

RUG Trigger

RUG Triggers were designed and installed at the 30-mile and 61-mile stations. The RUG Trigger (a separate RUGID Computer RTU from the one used to turn on the satellite-telemetry system) allows a YSI sonde to trigger an ISCO pump sampler and to connect to the satellite-telemetry system. Although turbidity is the parameter used by the RUG Trigger for reasons described previously, the RUG Trigger can be programmed to use any water-quality parameter measured by a YSI sonde. Without use of the RUG Trigger, the YSI sonde could connect directly to an ISCO, like the SDI–12 setup at the 225-mile station, or connect to a satellite-telemetry system, but not both. The ability to connect to both the ISCO pump sampler and the satellite-telemetry system is critical at the two more-remote stations with YSI sondes (30-mile and 61-mile). In addition, the RUG Trigger can be programmed to trigger the ISCO pump sampler after turbidity from multiple consecutive (usually two) measurements exceeds the set level. This prevents pump samples from being collected as a result of anomalous turbidity spikes.

The RUG Trigger controls the flow of communication from the YSI sonde. During normal operation, the YSI sonde, connected to the RUG Trigger, records water-quality measurements (typically water temperature, specific conductance, turbidity, and dissolved oxygen) every 15 minutes and is accessible, through the satellite-telemetry system, for monitoring and data downloading. During the interval when the YSI sonde is not measuring a water-quality parameter, the RUG Trigger temporarily takes control of YSI communication to prompt the YSI sonde to record a measurement. These data are not recorded on the YSI sonde datalogger but are used by the RUG Trigger to determine if the turbidity (or other user-specified water-quality parameter) exceeds a preprogrammed threshold. If two consecutive (or other user-specified consecutive number) turbidity measurements exceed the threshold, the RUG Trigger enables the ISCO sampling program. The ISCO pump sampler remains enabled and continues to run its program until turbidity from two consecutive measurements decreases to less than the set threshold.[5] Additional information, instructions, and a wiring-connection diagram for the RUG Trigger are given in appendix 3.

Discharge Calculations

Suspended sand, silt, and clay loads—not just concentrations—are required by the Glen Canyon Dam Adaptive Management Program to make decisions on dam operations. Because sediment loads are a product of suspended-sediment concentration and water discharge, the instantaneous discharge of water must be calculated at all monitoring stations that are not co-located with active USGS streamflow-gaging stations.

This is accomplished at these stations (30-mile, 61-mile, and 166-mile) by using standard USGS methods to convert stage measured by the ADPs to water discharge (Rantz and others, 1982). At each station, stage is measured every 15 minutes by the 1- and 2-MHz ADPs using an upward acoustic beam. The ADPs are designed such that these acoustic measurements of stage are automatically filtered to remove bad values by using simultaneous pressure measurements. The vendor-reported accuracy of these stage measurements is 3 millimeters (mm; identical to the 0.01-ft accuracy standard of stages measured at USGS streamflow-gaging stations); the precision of the ADP stage measurements is 1 mm. Discharge is measured periodically at the 30-mile, 61-mile, and 166-mile stations by using a boat-mounted acoustic-Doppler current profiler (ADCP) at measurement cross sections under temporary taglines. Six ADCP measurements are typically made over a reasonably wide range in stage (depending on the daily range in dam operations) at each of these stations twice or more per year. These measurement cross sections are also the locations at which EWI measurements are made to calibrate the acoustic, laser-diffraction, and pump-sampler instruments to measure the velocity-weighted suspended-sediment concentration and grain-size distribution in the adjacent river cross section. Following initial periods (up to several years) of calibration development, subsequent EWI measurements at these cross sections are used to verify these instrument calibrations. Eight to ten EWI measurements are thus typically made in association with the ADCP measurements at each of these stations twice or more per year.

ADCP measurements are collected by using RD Instruments Workhorse instruments. During the flow conditions that exist under typical dam operations, discharge is measured, following USGS procedures (Oberg and others, 2005; Mueller and Wagner, 2009), with a 600-kHz instrument. ADCPs are mounted on the side of motorized, aluminum, V-shaped-hull boats, and discharge is measured under a temporary tagline at the measurement cross sections (fig. 10). Edge coefficients are set as functions of the channel cross-section shape and cross-stream distribution of velocity near the channel margins at these measurement cross sections (Mueller and Wagner, 2009). To avoid the introduction of 2 to 3 percent biases in discharge measurements made by different ADCP operators, the edge coefficients are held constant for fixed ranges of discharge at each measurement cross section. On most rivers, Global Positioning System (GPS) measurements are used to measure the location of the boat independently from ADCP bottom tracking measurements, and thus detect and correct for moving-bed conditions in discharge measurements (Mueller and Wagner, 2009). The poor GPS coverage at the sediment-transport monitoring stations in Marble and Grand Canyons necessitated the use of loop tests to detect and correct for moving-bed conditions. Loop tests are conducted under the temporary tagline before each ADCP measurement to determine whether moving-bed conditions are present and, if moving-bed conditions are detected, are used to correct the discharge measurements to compensate for the effect of a moving bed. If moving-bed conditions are present and ADCP bottom tracking is uninterrupted, discharges

---

[5] Again, the user can specify the average of any number of measurements of any water-quality parameter to define this threshold.

are corrected by using Loop Correction software v1.6 (U.S. Geological Survey, 2008a). During controlled-flood events (November 2004 and March 2008), discharge measurements were typically made with either a 600- or 300-kHz instrument; lower-frequency instruments performed better during the controlled floods because their measurements were less affected by the greater acoustic attenuation arising from higher concentrations of suspended sediment, the deeper water, and the greater moving-bed conditions. Discharges during controlled floods released from the dam (as documented, for example, by Topping and others, 2010) were corrected for moving-bed conditions by using the Loop Correction software (if bottom tracking was uninterrupted), the methods of Gartner and Ganju (2007), or with the Stationary Moving-Bed Analysis software v.4.2 (U.S. Geological Survey, 2008b). Limitations on ADCP use during controlled floods include moving-bed conditions, decreased sampling resolution resulting from the use of lower-frequency instruments, lack of bottom track (Gartner and Ganju, 2007), and ADP stage-record problems (from both greater acoustic attenuation and mounting-bracket breakages during the floods).

Problems with the ADCP measurements made during controlled floods, arising mostly from loss of ensembles as a result of greater acoustic attenuation, poor bottom track, and greater moving bed, in combination with ADP stage-record problems and possibly real changes in hydraulic controls, result in greater scatter in ADCP-measured discharges during controlled floods (fig. 11). Although many factors can affect the accuracy of an ADCP measurement (Mueller, 2002; Oberg and Mueller, 2007), for the purposes of this report, the measured ADCP discharges are assumed to be within 5 percent of the actual discharge (Mueller, 2003; Oberg and Mueller, 2007). Error bars in figure 11 therefore reflect this assumed 5-percent accuracy.

ADCP measurements made over the course of multiple station visits, and over a range of discharges, were combined with the ADP stage data to define the stage-discharge relations for the 30-mile, 61-mile, and 166-mile stations depicted in figure 11. Subsequent ADCP measurements collected during semiannual river trips are used to verify the stage-discharge relation at each of these three stations (table 3) and, if necessary, to develop the gage-height adjustments needed to apply the shifting-control method (Rantz and others, 1982). Discharges for 15-minute intervals at the 30-mile, 61-mile, and 166-mile stations are computed on the basis of these stage-discharge relations and are posted on the World Wide Web at http://www.gcmrc.gov.

ADCP discharge measurements were paired with concurrent ADP stage values to construct the stage-discharge relations (rating curves) in figure 11. Linear interpolation of the 15-minute stage data was used to calculate stage when discharge measurement temporal midpoints did not coincide with the 15-minute stage-data-collection interval. ADP stage and discharge were plotted and a rating curve was developed for each site following the methods in Rantz and others (1982). Because of discharge regulation of the Colorado River by Glen Canyon Dam, the likely potential discharges are limited to a range of 142 to 1,275 m^3/s. Over this range, a second order polynomial was deemed sufficient (coefficient of determination, R^2, values greater than 0.99) to define the rating curves at the 30-mile, 61-mile, and 166-mile stations (fig. 11). Because measured

Figure 10. Aluminum V-shaped-hull motorized boat used to make ADCP and EWI measurements. During EWI measurements, the US D–96–A1 sampler is deployed from a crane mounted to the front of the boat.

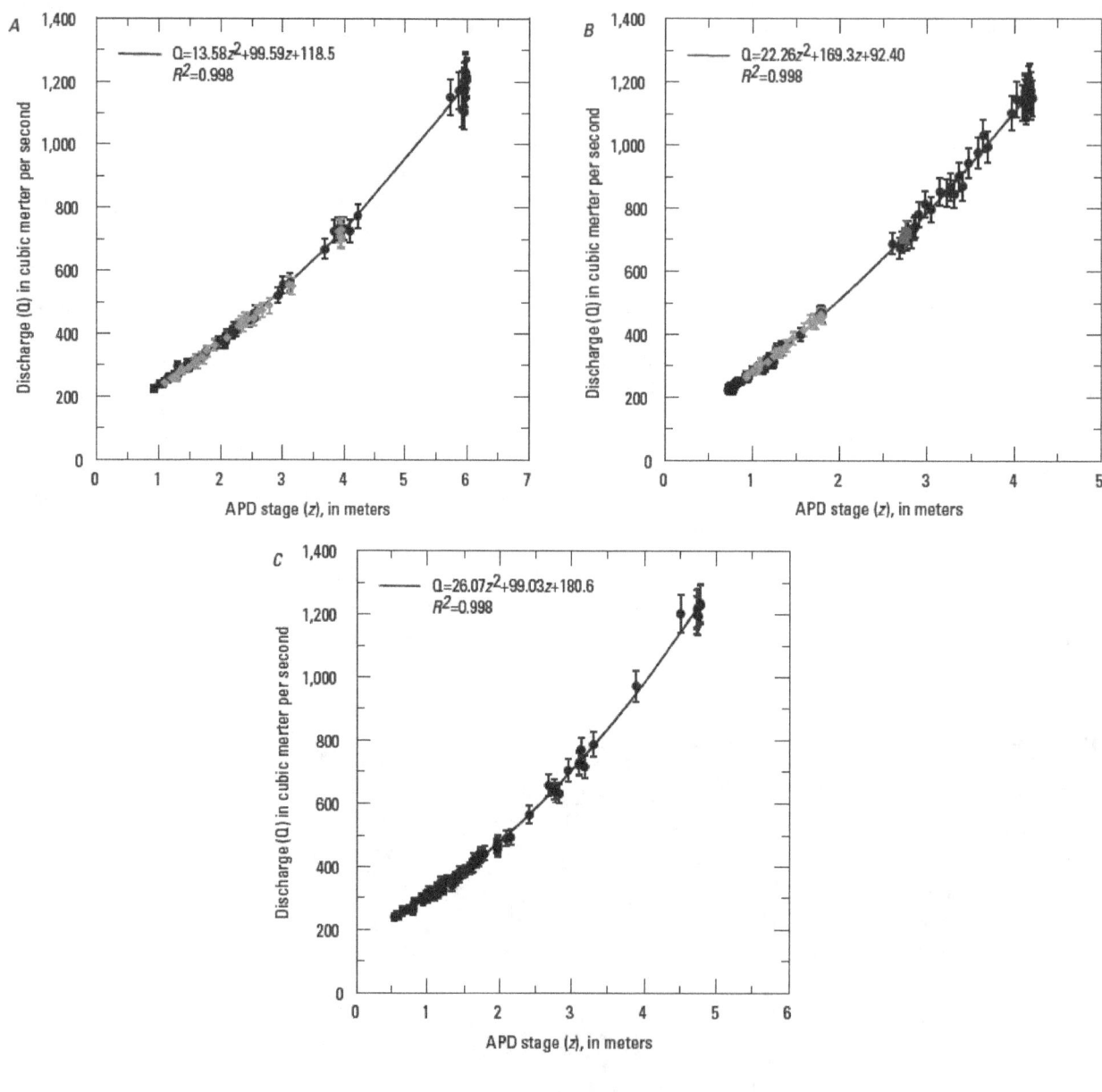

EXPLANATION

- ADCP calibration discharge measurements and bars indicating accuracy of ± 5 percent
- ADCP verification discharge measurements and bars indicating accuracy of ± 5 percent
- ADCP shift-adjusted verification discharge measurements from August 2011 and bars indicating accuracy of ± 5 percent

Figure 11. Stage-discharge relations at the sediment-transport monitoring stations. *A*, 30-mile station, where stage is measured by using the 1-MHz ADP deployed upstream from 30 Mile Rapid (see fig. 3*A*). *B*, 61-mile station, where stage is measured by the 1-MHz ADP (see fig. 3*B*). *C*, 166-mile station, where stage is measured by the 1-MHz ADP located at the former location of USGS streamflow-gaging station Colorado River above National Canyon near Supai, Arizona (09404120; see fig. 3*D*).

discharges cover most of the possible range in discharge at the stations, the rating-curve predicted discharges are not greatly extrapolated beyond the measured discharges. ADCP measurements made after the rating curves were established (August 2008 for the 30-mile and 61-mile stations, August 2011 for the 166-mile station) are used to verify the relations and to monitor for changes in the hydraulic control at each station (table 3).

In addition to using the ADCP-measured discharges on semiannual river trips to evaluate the continued accuracy of stage-discharge relations, a hydrographic-comparison method is also used to evaluate the continued accuracy of these relations and to help determine the times when hydraulic controls may have changed and resulted in shifts in the ratings. The hydrographic-comparison method incorporates known tributary and groundwater inflow and compares the rating-curve predicted discharges between the sediment-transport monitoring stations and the following active USGS streamflow-gaging stations: Colorado River at Lees Ferry, Ariz. (09380000); Paria River at Lees Ferry, Ariz. (09382000); Little Colorado River above mouth near Desert View, Ariz. (09402300); Colorado River near Grand Canyon, Ariz. (09402500); and Colorado River above Diamond Creek near Peach Springs, Ariz. (09404200). This method ensures that the principle of conservation of mass is not violated and helps identify stations where the rating curve has likely shifted. Because discharge can only be measured during semiannual river trips, it is important to verify that the computed discharges at the sediment-transport monitoring stations and the active USGS streamflow-gaging stations agree. If shifts in the stage-discharge relation are detected, either through the hydrographic-comparison method or by the observation of sustained systematic errors between the rating-curve predicted discharges and the measured discharges, the shifting-control method is used to develop and apply shifts to the ADP stage record (Rantz and others, 1982). In the three years since the rating curves where developed for the 30-mile and 61-mile stations the shifting-control method has never been used at the 30-mile station and has been employed only since May 2011 at the 61-mile station. Because discharge is measured typically six times twice a year at each sediment-transport monitoring station (table 3), single measurements exceeding a 5-percent departure from the rating curve are not considered a basis for applying a shift. If multiple measurements from a single trip indicate a systematic error in the discharge predicted by the rating curve, or the average departure from the rating curve exceeds 3 percent, rating-curve predicted discharges at the station will be compared with those at other sediment-transport monitoring stations and at streamflow-gaging stations using the hydrographic-comparison method, and a shift may be applied. Negative shifts arise from aggradation of the channel bed in the hydraulic control during tributary floods. These changes in hydraulic control are often short lived, and the hydraulic controls typically revert back to their initial condition during higher dam releases. Relatively infrequent larger tributary floods or debris flows may significantly alter the hydraulic control, requiring the development of a new rating curve, but this has not occurred during the 8-year duration of this study.

Station Visits and Maintenance

In the Colorado River case study, stations are visited every one to two months (87-mile and 225-mile), every three months (30-mile and 61-mile), or twice a year (166-mile) for regular maintenance of the instruments and pump samplers. USGS cableways present at the 87-mile and 225-mile stations enable monthly to bi-monthly collection of EDI suspended-sediment samples used to verify the ADP, LISST–100, and ISCO pump-sampler calibrations. In addition, semiannual river trips are used for major site maintenance, ADCP discharge measurements, and collection of EDI or EWI (depending on station) suspended-sediment samples used to verify the ADP, LISST–100, and ISCO pump-sampler calibrations. When made correctly, EDI and EWI suspended-sediment measurements are identical (U.S. Geological Survey, 2006) and have identical error (Topping and others, 2011). The frequency of the site visits depends on the instrumentation at the site, the problems found during weekly instrument checks using the satellite-telemetry system, and the accessibility of the site. Major maintenance is usually completed only during the semiannual river trips, with routine maintenance completed during other site visits. Routine station maintenance consists of the following:

- Site inspection.

- Checking pump samplers for samples and proper operation. Samples, if present, are removed from the sampler and labeled with the station name, ISCO name, sample number, date and time of collection, and water level inside the sample bottle.

- Measuring and recording the voltage of all 12-volt battery banks.

- Cleaning the LISST–100 (if present), using bleach and an ultrasonic cleaning bath.

- Checking and calibrating the turbidity probes.

- Ensuring that the ADP mount is free of debris, and the ADP transducers are sediment free (if the ADP location is accessible by hiking).

Site inspection consists of inspecting the site wiring for rodent or wear damage, inspecting the satellite dish and its guy wires (if applicable), checking the condition of all solar panels, and inspecting the instrument mounts and cables. Information from each station visit is recorded in a notebook for future reference. At sites with USGS cableways (87-mile and 225-mile), EDI suspended-sediment samples and bed-material samples are collected during routine station visits.

River trips are used to complete major station maintenance, to make EDI or EWI suspended-sediment measurements and collect bed-material samples at all stations, and to make ADCP discharge measurements at the sediment-transport monitoring stations not co-located with active USGS streamflow-gaging

Table 3. Stage-discharge relation (rating curve) verification measurements.

[Measurements from a 600-kHz acoustic Doppler current profiler. Magnitudes of applied shifts are based on these measurements and application of the hydrographic-comparison method.]

Station name	Measurement date	Measurement time (MST, 24 hour)	Measured discharge (m³/s)	Rating-curve predicted discharge (m³/s)	Percent error	Shift applied (m)
30-mile	08/22/2008	1516	344.2	340.3	-1.1	0
30-mile	08/22/2008	2034	467.6	470.4	0.59	0
30-mile	02/28/2009	0746	318.7	334.1	4.8	0
30-mile	02/28/2009	1040	256.7	272.1	6.0	0
30-mile	02/28/2009	1308	241.2	245.9	2.0	0
30-mile	08/21/2009	1028	315.7	314.3	-0.43	0
30-mile	08/21/2009	1304	272.6	275.2	0.97	0
30-mile	08/21/2009	1623	288.9	297.2	2.9	0
30-mile	08/21/2009	2031	427.2	433.9	1.6	0
30-mile	08/22/2009	0613	423.6	420.5	-0.73	0
30-mile	08/22/2009	1051	304.6	311.6	2.3	0
30-mile	02/19/2010	0922	300.0	309.2	3.0	0
30-mile	02/19/2010	1341	261.7	267.9	2.4	0
30-mile	02/19/2010	1935	386.8	389.5	0.72	0
30-mile	02/20/2010	0904	304.0	318.5	4.8	0
30-mile	02/20/2010	1223	256.3	260.6	1.7	0
30-mile	02/20/2010	1528	258.4	263.3	1.9	0
30-mile	08/19/2010	1540	286.1	282.1	-1.4	0
30-mile	08/19/2010	1801	362.1	360.2	-0.51	0
30-mile	08/20/2010	0631	434.8	425.7	-2.1	0
30-mile	08/20/2010	1033	324.1	328.8	1.5	0
30-mile	08/20/2010	1315	281.2	283.7	0.87	0
30-mile	08/20/2010	2007	443.6	434.4	-2.1	0
30-mile	02/20/2011	1534	488.7	503.8	3.1	0
30-mile	02/20/2011	1843	555.5	561.2	1.0	0
30-mile	02/20/2011	2041	551.8	571.2	3.5	0
30-mile	02/21/2011	0934	448.4	460.4	2.7	0
30-mile	02/21/2011	1212	445.2	444.6	-0.13	0
30-mile	02/21/2011	1434	474.0	485.1	2.4	0
30-mile	08/19/2011	0847	707.6	727.4	2.8	0
30-mile	08/19/2011	1319	707.5	724.5	2.4	0
30-mile	08/19/2011	1921	728.3	718.7	-1.3	0
30-mile	08/20/2011	0805	732.7	727.8	-0.66	0
30-mile	08/20/2011	1308	713.3	725.3	1.7	0
30-mile	08/20/2011	1839	711.5	723.9	1.7	0

Table 3. Stage-discharge relation (rating curve) verification measurements.—Continued

Station name	Measurement date	Measurement time (MST, 24 hour)	Measured discharge (m³/s)	Rating-curve predicted discharge (m³/s)	Percent error	Shift applied (m)
61-mile	08/25/2008	1039	463.4	470.0	1.4	0
61-mile	08/25/2008	1822	350.8	345.6	-1.5	0
61-mile	03/03/2009	1143	351.5	350.1	-0.39	0
61-mile	03/03/2009	1330	336.2	337.2	0.30	0
61-mile	03/03/2009	1639	297.5	293.6	-1.3	0
61-mile	08/24/2009	0656	420.5	416.7	-0.91	0
61-mile	08/24/2009	1321	394.7	387.0	-1.9	0
61-mile	08/24/2009	1521	357.2	351.2	-1.7	0
61-mile	08/24/2009	1814	314.9	303.0	-3.8	0
61-mile	08/24/2009	2131	276.9	271.8	-1.8	0
61-mile	08/25/2009	700	468.5	462.5	-1.3	0
61-mile	02/22/2010	1026	295.2	298.8	1.2	0
61-mile	02/22/2010	1304	319.2	321.0	0.56	0
61-mile	02/22/2010	1734	289.2	284.8	-1.5	0
61-mile	02/22/2010	2043	268.6	270.2	0.56	0
61-mile	02/23/2010	0719	347.8	353.8	1.7	0
61-mile	02/23/2010	0858	356.7	359.9	0.90	0
61-mile	02/23/2010	1058	368.6	371.4	0.77	0
61-mile	02/23/2010	1227	370.0	369.1	-0.23	0
61-mile	08/22/2010	1630	352.5	340.8	-3.3	0
61-mile	08/22/2010	1837	313.8	304.5	-3.0	0
61-mile	08/23/2010	0645	448.4	433.6	-3.3	0
61-mile	08/23/2010	0838	449.3	444.1	-1.2	0
61-mile	08/24/2010	0738	470.3	464.0	-1.3	0
61-mile	08/24/2010	0929	464.7	461.8	-0.61	0
61-mile	02/23/2011	1009	446.9	449.9	0.68	0
61-mile	02/23/2011	2010	450.2	446.3	-0.88	0
61-mile	02/24/2011	0822	447.0	445.8	-0.27	0
61-mile	02/24/2011	1205	451.0	446.3	-1.1	0
61-mile	02/24/2011	1419	445.6	445.3	-0.08	0
61-mile	02/24/2011	1556	446.0	446.3	0.06	0
61-mile	08/22/2011	0822	715.3	720.5	0.72	0.07
61-mile	08/22/2011	1329	733.3	725.7	-1.0	0.07
61-mile	08/22/2011	1946	736.5	723.4	-1.8	0.07
61-mile	08/23/2011	0808	714.9	719.9	0.70	0.07
61-mile	08/23/2011	1340	734.9	723.7	-1.5	0.07
61-mile	08/23/2011	1854	731.8	728.1	-0.52	0.07

stations. Station maintenance completed during river trips is similar to that done during the routine site visits; however, in addition to standard maintenance, major maintenance requiring more time or equipment is completed. Semiannual river trips follow the same general work plan. An example of a typical work plan for a river trip is given in appendix 4.

Summary

To meet the needs of resource managers in the Department of the Interior's Glen Canyon Dam Adaptive Management Program, the USGS Grand Canyon Monitoring and Research Center constructed and tested a sediment-transport monitoring network for the Colorado River within Marble and Grand Canyons in Grand Canyon National Park. Because of large discharge-independent changes in suspended-sediment concentration and grain size that can occur over timescales of less than an hour, high-temporal-resolution data are required for accurate calculations of sediment transport in this river. The monitoring network is therefore used to collect 15-minute-resolution suspended-sediment data using acoustic and laser-diffraction measurements at five stations on the Colorado River in the study area. Three of the five monitoring stations have two-way satellite-telemetry systems that provide biweekly updates of Web-served data; data from the other two sites are updated following site visits. The two-way satellite-telemetry systems allow remote communication with any instrument by using the vendor-supplied software for that instrument. Though developed for the Colorado River, the sediment-transport monitoring network described in this report could be used on any river to facilitate the collection and near real-time processing of suspended-sediment data. The techniques and methods described herein can easily be applied to other rivers and are especially appropriate for remote locations where site access is difficult.

Acknowledgments

This project was funded by the U.S. Department of the Interior's Glen Canyon Dam Adaptive Management Program through the USGS Grand Canyon Monitoring and Research Center. Scott Wright (USGS California Water Science Center) was instrumental in making improvements in the design of the monitoring network as well as making fundamental advancements in the processing of the acoustic data collected in this network. Jim Eads (USGS Utah Water Science Center) helped develop the use of broadband satellite telemetry in remote field locations. Cory Lochridge (USGS Grand Canyon Monitoring and Research Center) assisted in the programming of the SatControl program used in the satellite-telemetry systems.

References Cited

Agrawal, Y.C., and Pottsmith, H.C., 1994, Laser diffraction particle sizing in STRESS: Continental Shelf Research, v. 14, no. 10/11, p. 1,101–1,121.

Agrawal, Y.C., and Pottsmith, H.C., 2000, Instruments for particle size and settling velocity observations in sediment transport: Marine Geology, v. 168, p. 89–114.

Campbell, David, Durst, Scott, Kantola, A.T., Kubly, D.M., Muth, R.T., Swett, John, and Whitmore, Sharon, 2010, Overview of the Colorado River Basin collaborative management programs, in Melis, T.S., Hamill, J.F., Bennett, G.E., Coggins, L.G., Jr., Grams, P.E., Kennedy, T.A., Kubly, D.M., and Ralston, B.E. eds., Proceedings of the Colorado River Basin Science and Resource Management Symposium, November 18–20, 2008, Scottsdale, Arizona: U.S. Geological Survey Scientific Investigations Report 2010–5135, p.15–42. (Also available at http://pubs.usgs. gov/sir/2010/5135/.)

Davis, B.E, 2001, Report PP, The US D–96—an isokinetic suspended-sediment/water quality collapsible bag sampler: Vicksburg, Miss., Federal Interagency Sedimentation Project, 37 p. (Also available at http://water.usgs.gov/ fisp/docs/Report_PP,_US_D-96_011114_.pdf.)

Dolan, R., Howard, A., and Gallenson, A., 1974, Man's impact on the Colorado River in the Grand Canyon: American Scientist, v. 62, p. 392–401.

Edwards, T.K., and Glysson, G.D., 1999, Field methods for measurement of fluvial sediment: Techniques of Water-Resources Investigations of the U.S. Geological Survey, book 3, chap. C2, 89 p. (Also available at http://pubs.usgs. gov/twri/twri3-c2/)

Federal Interagency Sedimentation Project, 2003, Report PP, The US D–96—an isokinetic suspended-sediment/water quality collapsible bag sampler, addendum – II, The US D–96–A1—a lightweight version of the US–96: Vicksburg, Mississippi, Federal Interagency Sedimentation Project, 4 p. (Also available at http://water.usgs.gov/fisp/docs/ Report_PP-Addendum_II,_US_D-96_030507_.pdf.)

Garrett, W.B., Van De Vanter, E.K., and Graf, J.B., 1993, Streamflow and sediment-transport data, Colorado River and three tributaries in Grand Canyon, Arizona, 1983 and 1985-86: U.S. Geologic Survey Open-File Report 93-174, 624 p. (Also available at http://pubs.usgs.gov/ of/1993/0174/report.pdf.)

Gartner, J.W., and Ganju, N.K., 2007, Correcting acoustic Doppler current profiler discharge measurements bias from moving-bed conditions without global positioning during the 2004 Glen Canyon Dam controlled flood on the Colorado River: Limnology and Oceanography: Methods, v. 5, p.156–162.

Grant, G.E., Schmidt, J.C., and Lewis, S.L., 2003, A geological framework for interpreting downstream effects of dams on rivers, in O'Connor, J.E., and Grant, G. E., eds., A peculiar river—geology, geomorphology, and hydrology of the Deschutes River, Oregon: Washington, D.C., American Geophysical Union Water Science and Application, v. 7, p. 203–219.

Gray, J.R., and Gartner, J.W., 2009, Technological advances in suspended-sediment surrogate monitoring: Water Resources Research, v. 45, W00D29, 20 p., doi:10.1029/2008WR007063. (Also available at http://water.usgs.gov/osw/techniques/2008WR007063.pdf.)

Gray, J.R., Patino, Eduardo, Rasmussen, P.R., Larsen, M.C., Melis, T.S., Topping, D.J., Runner, M.S., and Alamo, C.F., 2003. Evaluation of sediment-surrogate technologies for computation of suspended-sediment transport, *Proceedings of the 1st International Yellow River Forum on River Basin Management*, the Yellow River Conservancy Publishing House, Zhengzhou, Henan Province, China, v. III, p. 314–323. (Also available at http://water.usgs.gov/osw/techniques/yrcc_surrogates.pdf.)

Greig, S.M., Sear, D.A., and Carling, P.A., 2005, The impact of fine sediment accumulation on the survival of incubating salmon progeny—implications for sediment management: Science of the Total Environment, v. 344, p. 241–258. (Also available at http://www.sciencedirect.com/science/article/pii/S004896970500118X

Griffiths, R.E., Topping, D.J., McDonald, R.R., and Sabol, T.A., 2010, The use of the multidimensional surface-water modeling system (MD_SWMS) in calculating discharge and sediment transport in remote ephemeral streams: Proceedings of the Joint Federal Interagency Conference on Sedimentation and Hydrologic Modeling, June 27–July 1, 2010, Las Vegas, Nev.

Guy, H.P., 1964, An analysis of some storm-period variables affecting stream sediment transport: U.S. Geological Survey Professional Paper 462–E, 46 p. (Also available at http://pubs.usgs.gov/pp/0462e/.)

Guy, H.P., 1970, Fluvial sediment concepts: U.S. Geological Survey Techniques of Water-Resources Investigations, book 3, chap. C1, 55 p. (Also available at http://pubs.usgs.gov/twri/twri3-c1/.)

Hazel, J.E., Jr., Grams, P.E., Schmidt, J.C., and Kaplinski, Matt, 2010, Sandbar response in Marble and Grand Canyons, Arizona, following the 2008 high-flow experiment on the Colorado River: U.S. Geological Survey Scientific Investigations Report 2010–5015, 52 p. (Also available at http://pubs.usgs.gov/sir/2010/5015/.)

Larsen, M.C., Gellis, A.C., Glysson, G.D., Gray, J.R., and Horowitz, A.J., 2010, Fluvial sediment in the environment—a national challenge: Proceedings of the Joint Federal Interagency Conference on Sedimentation and Hydrologic Modeling, June 27–July 1, 2010, Las Vegas, Nev.

Lisle, T.E., and Napolitano, M.B., 1998, Effects of recent logging on the main channel of North Fork Caspar Creek: United States Department of Agriculture Forest Service General Technical Report PSW–GTR–168, p. 81–85. (Also available at http://treesearch fs fed.us/pubs/7865.)

Love, S.K., 1954, Quality of surface waters of the United States, 1950, Parts 9-14: Colorado River basin to Pacific slope basins in Oregon and lower Columbia River basin: U.S. Geological Survey Water-Supply Paper 1189, 158 p. (Also available at http://pubs.usgs.gov/wsp/1189/report.pdf.)

Love, S.K., 1955, Quality of surface waters of the United States, 1951, Parts 9-14: Colorado River basin to Pacific slope basins in Oregon and lower Columbia River basin: U.S. Geological Survey Water-Supply Paper 1200, 285 p. (Also available at http://pubs.usgs.gov/wsp/1200/report.pdf.)

Love, S.K., 1957, Quality of surface waters of the United States, 1952, Parts 9-14: Colorado River basin to Pacific slope basins in Oregon and lower Columbia River basin: U.S. Geological Survey Water-Supply Paper 1253, 344 p. (Also available at http://pubs.usgs.gov/wsp/1253/report.pdf.)

Love, S.K., 1958, Quality of surface waters of the United States, 1953, Parts 9-14: Colorado River basin to Pacific slope basins in Oregon and lower Columbia River basin: U.S. Geological Survey Water-Supply Paper 1293, 372 p. (Also available at http://pubs.usgs.gov/wsp/1293/report.pdf.)

Love, S.K., 1959a, Quality of surface waters of the United States, 1954, Parts 9-14: Colorado River basin to Pacific slope basins in Oregon and lower Columbia River basin: U.S. Geological Survey Water-Supply Paper 1353, 426 p. (Also available at http://pubs.usgs.gov/wsp/1353/report.pdf.)

Love, S.K., 1959b, Quality of surface waters of the United States, 1955, Parts 9-14: Colorado River basin to Pacific slope basins in Oregon and lower Columbia River basin: U.S. Geological Survey Water-Supply Paper 1403, 437 p. (Also available at http://pubs.usgs.gov/wsp/1403/report.pdf).

Love, S.K., 1960, Quality of surface waters of the United States, 1956, Parts 9-14: Colorado River basin to Pacific slope basins in Oregon and lower Columbia River basin: U.S. Geological Survey Water-Supply Paper 1453, 447 p. (Also available at http://pubs.usgs.gov/wsp/1453/report.pdf.)

Love, S.K., 1961, Quality of surface waters of the United States, 1957, Parts 9-14: Colorado River basin to Pacific slope basins in Oregon and lower Columbia River basin: U.S. Geological Survey Water-Supply Paper 1523, 497 p. (Also available at http://pubs.usgs.gov/wsp/1523/report.pdf.)

Love, S.K., 1964a, Quality of surface waters of the United States, 1958, Parts 9-14: Colorado River basin to Pacific slope basins in Oregon and lower Columbia River basin: U.S. Geological Survey Water-Supply Paper 1574, 487 p. (Also available at http://pubs.usgs.gov/wsp/1574/report.pdf.)

Love, S.K., 1964b, Quality of surface waters of the United States, 1962, Parts 9-14: Colorado River basin to Pacific slope basins in Oregon and lower Columbia River basin: U.S. Geological Survey Water-Supply Paper 1945, 691 p. (Also available at http://pubs.usgs.gov/wsp/1945/report.pdf.)

Love, S.K., 1966, Quality of surface waters of the United States, 1959, Parts 9-14: Colorado River basin to Pacific slope basins in Oregon and lower Columbia River basin: U.S. Geological Survey Water-Supply Paper 1645, 524 p. (Also available at http://pubs.usgs.gov/wsp/1645/report.pdf.)

Madej, M.A., and Ozaki, V., 1996, Channel response to sediment wave propagation and movement, Redwood Creek, California, USA: Earth Surface Processes and Landforms, v. 21, p. 911–927. (Also available at http://onlinelibrary.wiley.com/doi/10.1002/(SICI)1096-9837(199610)21:10%3C911::AID-ESP621%3E3.0.CO;2-1/pdf.)

Melis, T.S., Topping, D.J., and Rubin D.M., 2003, Testing laser-based sensors for continuous in situ monitoring of suspended sediment in the Colorado River, Arizona, in Bogen, J., Fergus, T., and Walling, D.E., eds., Erosion and Sediment Transport Measurement in Rivers—Technological and Methodological Advances: Wallingford, Oxfordshire, United Kingdom, IAHS Press, IAHS Publication 283, p. 21–27.

Melis, T.S., Topping, D.J., Rubin, D.M, and Wright. S.A., 2007, Research furthers conservation of Grand Canyon sandbars: U.S. Geological Survey Fact Sheet 2007–3020, 4 p. (Also available at http://pubs.usgs.gov/fs/2007/3020/.)

Mueller, D.S., 2002, Field assessment of acoustic Doppler based discharge measurements, in Wahl, T.L., Pugh, C.A., Oberg, K.A., and Vermeyen, T.B., eds., Hydraulic measurements and experimental methods 2002, Proceedings of the Specialty Conference, July 28–August 1, 2002, Estes Park, Colo.: American Society of Civil Engineers, Reston, Va. (CD–ROM; Also available at http://cedb.asce.org/cgi/WWWdisplay.cgi?134128

Mueller, D.S., 2003, Field evaluation of boat-mounted acoustic Doppler instruments used to measure streamflow, in Rizoli, J.A., ed., Proceedings of the IEEE/OES Seventh Working Conference on Current Measurement Technology, March 13–15, 2003, San Diego, Calif., Sponsored by the Current Measurement Technology Committee of the Oceanic Engineering Society and the Institute of Electrical and Electronics Engineers, Piscataway, N.J. (Also available at http://ieeexplore.ieee.org/xpl/freeabs_all.jsp?arnumber=1194278&isnumber=26861; http://hydroacoustics.usgs.gov/publications/CMTC_Paper_David_S_Mueller.pdf.)

Mueller, D.S., and Wagner, C.R., 2009, Measuring discharge with acoustic Doppler current profilers from a moving boat: U.S. Geological Survey Techniques and Methods 3A–22, 72 p. (Also available at http://pubs.usgs.gov/tm/3a22.)

Nordin, C.F., Jr., and Beverage, J.P., 1965, Sediment transport in the Rio Grande, New Mexico: U.S. Geological Survey Professional Paper 462–F, 35 p. (Also available at http://pubs.usgs.gov/pp/0462f/.)

Oberg, K.A., Morlock, S.E., and Caldwell, W.S., 2005, Quality-assurance plan for discharge measurements using acoustic Doppler current profilers: U.S. Geological Survey Scientific Investigations Report 2005–5183, 35 p. (Also available at http://pubs.usgs.gov/sir/2005/5183/.)

Oberg, K.A., and Mueller, D.S., 2007, Validation of streamflow measurements made with acoustic Doppler current profilers: Journal of Hydraulic Engineering, v. 133, no. 12, p. 1,421–1,432. (Also available at http://hydroacoustics.usgs.gov/publications/14-Oberg-Mueller.pdf.)

Owens, P.N., Batalla, R.J., Collins, A.J., Gomez, B., Hicks, D.M., Horowitz, A.J., Kondolf, G.M., Marden, M., Page, M.J., Peacock, D.H., Petticrew, E.L., Salomons, W., and Trustrum, N.A., 2005, Fine-grained sediment in river systems—environmental significance and management issues: River Research and Applications, v. 21, p. 693–717, doi:10.1002/rra.878. (Also available at http://onlinelibrary.wiley.com/doi/10.1002/rra.878/pdf.)

Paulsen, C.G., 1949, Quality of surface waters of the United States, 1945: U.S. Geological Survey Water-Supply Paper 1030, 335 p. (Also available at http://pubs.usgs.gov/wsp/1030/report.pdf.)

Paulsen, C.G., 1950, Quality of surface waters of the United States, 1946: U.S. Geological Survey Water-Supply Paper 1050, 486 p. (Also available at http://pubs.usgs.gov/wsp/1050/report.pdf.)

Paulsen, C.G., 1952a, Quality of surface waters of the United States, 1947: U.S. Geological Survey Water-Supply Paper 1102, 651 p. (Also available at http://pubs.usgs.gov/wsp/1102/report.pdf.)

Paulsen, C.G., 1952b, Quality of surface waters of the United States, 1948, Parts 7-14: U.S. Geological Survey Water-Supply Paper 1133, 373 p. (Also available at http://pubs.usgs.gov/wsp/1133/report.pdf.)

Paulsen, C.G., 1953, Quality of surface waters of the United States, 1949, Parts 7-14: U.S. Geological Survey Water-Supply Paper 1163, 504 p. (Also available http://pubs.usgs.gov/wsp/1163/report.pdf.)

Porterfield, George, 1972, Computation of fluvial sediment discharge: U.S. Geological Survey Techniques of Water-Resources Investigations, book 3, chap. C3, 66 p. (Also available at http://pubs.usgs.gov/twri/twri3-c3/.)

Randle, T.J., and Bountry, J.A., 2010, Elwah River restoration—sediment adaptive management: Proceedings of the Joint Federal Interagency Conference on Sedimentation and Hydrologic Modeling, June 27–July 1, 2010, Las Vegas, Nev.

Rantz, S.E., and others, 1982, Measurement and computation of streamflow—Volume 2, Computation of discharge: U.S. Geological Survey Water-Supply Paper 2175, p. 285–631. (Also available at http://pubs.usgs.gov/wsp/wsp2175.)

Rubin, D.M., and Topping, D.J., 2001, Quantifying the relative importance of flow regulation and grain-size regulation of suspended-sediment transport (α) and tracking changes in bed-sediment grain size (β): Water Resources Research, v. 37, p. 133–146.

Rubin, D.M., and Topping, D.J., 2008, Correction to "Quantifying the relative importance of flow regulation and grain-size regulation of suspended-sediment transport (α) and tracking changes in bed-sediment grain size (β)": Water Resources Research, v. 44, W09701, 5 p., doi:10.1029/2008WR006819. (Also available at http://www.agu.org/journals/wr/wr0809/20 08WR006819/2008WR006819.pdf.)

Rubin, D.M., Topping, D.J., Schmidt, J.C., Hazel, J., Kaplinski, M., and Melis, T.S., 2002, Recent sediment studies refute Glen Canyon Dam hypothesis: EOS, Transactions, American Geophysical Union, v. 83, no. 25, p. 273, 277–278.

Schmidt, J.C., Topping, D.J., Grams, P.E., and Hazel, J.E., 2004, System-wide changes in the distribution of fine sediment in the Colorado River corridor between Glen Canyon Dam and Bright Angel Creek, Arizona: Final report to the USGS Grand Canyon Monitoring and Research Center, Flagstaff, Arizona, 107 p. (Also available at http://www.gcmrc.gov/library/reports/Physical/Fine_Sed/Schmidt2004.pdf.)

Schmidt, J.C., and Wilcock, P.R., 2008, Metrics for assessing the downstream effects of dams: Water Resources Research, v. 44, 19 p., W04404, doi:10.1029/2006WR005092. (Also available at http://www.agu.org/journals/wr/wr0804/2006W R005092/2006WR005092.pdf.)

Scott, C.H., and Stephens, H.D., 1966, Special sediment investigations–Mississippi River at St. Louis, Missouri, 1961-63: U.S. Geological Survey Water-Supply Paper 1819-J, 35 p. and 3 plates. (Also available at http://pubs.usgs.gov/wsp/1819j/report.pdf.)

Topping, D.J., Melis, T.S., Rubin, D.M., and Wright, S.A., 2004, High-resolution monitoring of suspended-sediment concentration and grain size in the Colorado River in Grand Canyon using a laser-acoustic system, in Hu, C., and Tan, Y., Eds., International Symposium on River Sedimentation, 9th, Yichang, People's Republic of China 2004, Proceedings, p. 2507-2514 (Also available at http://www.gcmrc.gov/library/reports/Physical/Fine_Sed/Topping2004.pdf.)

Topping, D.J., Rubin, D.M., Grams, P.E., Griffiths, R.E., Sabol, T.A., Voichick, N., Tusso, R.B., Vanaman, K.M., and McDonald, R.R., 2010, Sediment transport during three controlled-flood experiments on the Colorado River downstream from Glen Canyon Dam, with implications for eddy-sandbar deposition in Grand Canyon National Park: U.S. Geological Survey Open-File Report 2010–1128, 111 p. (Also available at http://pubs.usgs.gov/of/2010/1128.)

Topping, D.J., Rubin, D.M., and Melis, T.S., 2007, Coupled changes in sand grain size and sand transport driven by changes in the upstream supply of sand in the Colorado River; relative importance of changes in bed-sand grain size and bed-sand area: Sedimentary Geology, v. 202, p. 538–561, doi: 10.1016/j.sedgeo.2007.03.016. (Also available at http://dx.doi.org/10.1016/j.sedgeo.2007.03.016.)

Topping, D.J., Rubin, D.M., Nelson, J.M., Kinzel, III, P.J., and Bennett, J.P., 1999, Linkage between grain-size evolution and sediment depletion during Colorado River floods, in Webb, R.H., Schmidt, J.C., Marzolf, G.R., and Valdez, R.A., eds., The 1996 controlled flood in Grand Canyon: Washington, D.C., American Geophysical Union, Geophysical Monograph 110, p. 71-98.

Topping, D.J., Rubin, D.M., Nelson, J.M., Kinzel, P.J., III, and Corson, I.C., 2000, Colorado River sediment transport—Part 2, Systematic bed-elevation and grain-size effects of sand supply limitation: Water Resources Research, v. 36, p. 543–570. (Also available at http://www.agu.org/journals/wr/v036/i002/1999WR900286/1999WR900286.pdf.)

Topping, D.J., Rubin, D.M., Schmidt, J.C., Hazel, J.E., Jr., Melis, T.S., Wright, S.A., Kaplinski, M., Draut, A.E., and Breedlove, M.J., 2006, Comparison of sediment-transport and bar-response results from the 1996 and 2004 controlled-flood experiments on the Colorado River in Grand Canyon: Proceedings of the 8th Federal Inter-Agency Sedimentation Conference, Reno, Nev., April 2–6, 2006 (CD–ROM), ISBN 0-9779007-1-1. (Also available at http://pubs.usgs.gov/misc/FISC_1947-2006/pdf/1st-7thFISCs-CD/8thFISC/Session%201B- 3_Topping.pdf.)

Topping, D.J., Rubin, D.M., and Vierra, L.E., Jr., 2000, Colorado River sediment transport—Part 1, Natural sediment supply limitation and the influence of Glen Canyon Dam: Water Resources Research, v. 36, p. 515–542. (Also available at http://www.agu.org/journals/wr/v036/i002/1999WR900285/1999WR900285.pdf.)

Topping, D.J., Rubin, D.M., Wright, S.A., and Melis, T.S., 2011, Field evaluation of the error arising from inadequate time averaging in the standard use of depth-integrating suspended-sediment samplers: U.S. Geological Survey Professional Paper 1774.

Topping, D.J., Schmidt, J.C., and Vierra, L.E., Jr., 2003, Computation and analysis of the instantaneous-discharge record for the Colorado River at Lees Ferry, Arizona—May 8, 1921, through September 30, 2000: U.S. Geological Survey Professional Paper 1677, 118 p. (Also available at http://pubs.usgs.gov/pp/pp1677/)

Topping, D.J., Wright, S.A., Melis, T.S., and Rubin, D.M., 2006, High-resolution monitoring of suspended-sediment concentration and grain size in the Colorado River using laser-diffraction instruments and a three-frequency acoustic system: Proceedings of the 8th Federal Inter-Agency Sedimentation Conference, Reno, Nev., April 2–6, 2006 (CD–ROM), ISBN 0-9779007-1-1. (Also available at http://pubs.usgs.gov/misc_reports/FISC_1947-2006/pdf/1st-7thFISCs-CD/8thFISC/Session%206C-3_Topping.pdf.)

Topping, D.J., Wright, S.A., Melis, T.S., and Rubin, D.M., 2007, High-resolution measurements of suspended-sediment concentration and grain size in the Colorado River in Grand Canyon using a multi-frequency acoustic system: Proceedings of the Tenth International Symposium on River Sedimentation, August 1–4, 2007, Moscow, Russia, v. 3, p. 330–339. ISBN 978-5-89575-124-4, 978-5-89575-127-5. (Also available at http://www.gcmrc.gov/library/reports/physical/Fine_Sed/Topping2007b.pdf.)

U.S. Department of the Interior, 1995, Operation of Glen Canyon Dam, Final Environmental Impact Statement: Salt Lake City, Utah, Bureau of Reclamation, 337 pp.

U.S. Environmental Protection Agency, 2004, The incidence and severity of sediment contamination in surface waters of the United States, National Sediment Quality Survey (2d ed.): Washington, D.C., EPA–823–R–04–007 [variously paged].

U.S. Geological Survey, 1947, Quality of surface waters of the United States, 1944: U.S. Geological Survey Water-Supply Paper 1022, 311 p. (Also available at http://pubs.usgs.gov/wsp/1022/report.pdf.)

U.S. Geological Survey, 1972, Quality of surface waters of the United States, 1967, Parts 9-11: Colorado River basin to Pacific slope basins in California: U.S. Geological Survey Water-Supply Paper 2015, 702 p. (Also available at http://pubs.usgs.gov/wsp/2015/report.pdf.)

U.S. Geological Survey, 1976, Quality of surface waters of the United States, 1970, Parts 9 and 10: Colorado River basin and The Great Basin: U.S. Geological Survey Water-Supply Paper 2158, 371 p. (Also available at http://pubs.usgs.gov/wsp/2158/report.pdf.)

U.S. Geological Survey, 2006, Collection of Water Samples (ver. 2.0): U.S. Geological Survey Techniques of Water-Resources Investigations, book 9, chap. A4, accessed April 15, 2009, at http://pubs.water.usgs.gov/twri9A4.

U.S. Geological Survey, 2008a, Loop Correction software (LC): U.S. Geological Survey Web page, accessed March 4, 2008, at http://hydroacoustics.usgs.gov/movingboat/LC1.shtml.

U.S. Geological Survey, 2008b, Stationary Moving-Bed Analysis (SMBA): U.S. Geological Survey Web page, accessed March 4, 2008, at http://hydroacoustics.usgs.gov/movingboat/SMBA1.shtml.

U.S. Geological Survey, 2011a, Water quality samples for the nation, USGS, 09402500, Colorado River near Grand Canyon, Arizona: accessed on August 1, 2011, at http://nwis.waterdata.usgs.gov/nwis/qwdata/?site_no=09402500&agency_cd=USGS&.

U.S. Geological Survey, 2011b, Water quality samples for the nation, USGS, 07010000, Mississippi River at St. Louis, Missouri: accessed on August 1, 2011, at http://nwis.waterdata.usgs.gov/nwis/qwdata/?site_no=07010000&agency_cd=USGS&.

Wood, P.J., and Armitage, P.D., 1997, Biological effects of fine sediment in the lotic environment: Environmental Management, v. 21, no. 2, p. 203–217. (Also available at http://www.springerlink.com/content/ydag5vfcrtbkph4d/fulltext.pdf.)

Wright, S.A., Melis, T.S., Topping, D.J., and Rubin, D.M., 2005, Influence of Glen Canyon Dam operations on downstream sand resources of the Colorado River in Grand Canyon, in Gloss, S.P., Lovich, J.E., and Melis, T.S., eds., The state of the Colorado River ecosystem in Grand Canyon: U.S. Geological Survey Circular 1282, p. 17–31.

Wright, S.A., Schmidt, J.C., Melis, T.S., Topping, D.J., and Rubin, D.M., 2008, Is there enough sand? Evaluating the fate of Grand Canyon sandbars: GSA Today, v. 18, no. 8, p. 4–10, doi:10.1130/GSATG12A.1.

Wright, S.A., Topping, D.J., and Williams, C.A., 2010b, Discriminating silt-and-clay from suspended-sand in rivers using side-looking acoustic profilers: Proceedings of the Joint Federal Interagency Conference on Sedimentation and Hydrologic Modeling, June 27–July 1, 2010, Las Vegas, Nev. (Also available at http://acwi.gov/sos/pubs/2ndJFIC/Contents/2C_Wright_03_01_10_paper.pdf.)

Appendixes 1–5

Appendix 1: Acoustic-Doppler Profiler Mounting and Maintenance

Mounting Instructions

Acoustic-Doppler side-looking profilers (ADPs) need to be mounted level; the Nortek EasyQ and Aquadopp ADPs, and OTT SLD ADPs have internal pitch and roll sensors to help with this. The square root of the sum of the pitch squared plus the roll squared must be less than 3 degrees for the instrument to function properly. To adjust the ADP pitch and roll:

(1) Mount the ADP in the water as close to horizontal as possible

(2) Connect to the instrument with the appropriate vendor-supplied software and select "On-Line"→ "Start Range Check."

(3) Note the pitch and roll vales. If the pitch and roll values are within the standards, finish securing the ADP mount; if they are not, proceed with step 4.

   o   Pitch – rotation around the long axis
   Negative numbers mean that the horizontal acoustic beams (beams 1 and 2) are pointed up towards the river surface.
   Positive numbers mean that the horizontal acoustic beams (beams 1 and 2) are pointed down towards the river bed.

   o   Roll – rotation around the short axis
   Negative numbers mean that the sensor head is pointed towards the riverbed.
   Positive numbers mean that the sensor head is pointed towards the water surface.

(4) Remove the ADP mount from the water and position it so that it is oriented close to its orientation while in the water.

(5) Note the pitch and roll values with the mount out of the water and calculate the necessary out-of-water pitch and roll values (for example, if the pitch was -2 degrees in the water and 8 degrees out of the water, the ADP needs to be rotated positive 2 degrees so that the out-of-water pitch value is 10 degrees).

(6) Reposition the ADP on the mount by the calculated values.

(7) Remount the ADP in the water and repeat steps 3–7 as necessary.

Maintenance and Troubleshooting

ADP maintenance includes cleaning the transducers of built up sediment and algal growth, calibrating the temperature sensor, cleaning the pressure transducer, and resetting the pressure offset, if necessary. After cleaning the transducers, a range check using the vendor-supplied software is conducted to ensure that the acoustic measurements made along the various beams agree. The temperature sensor in the ADP must be checked and, if necessary, calibrated to ensure accurate calculations of the speed of sound needed for cell positioning. The drift in the temperature sensors in the Nortek and OTT ADPs has been found to be less than several tenths of a degree over 6 months in the Colorado River, so calibration of the temperature sensor is typically needed only once or twice per year. The temperature-calibration procedure is as follows:

(1) Download any data files needed from the instrument; the data files are erased during the calibration process.

(2) Load the instrument deployment file.

(3) Complete a Stage or Range check (temperature calibration will not work without this step).
   o   Select "On-Line"→ "Start Stage (or Range) Check."
   o   Select "Stop Recorder Deployment."

(4) Select "Temperature Calibration."
   o   Select "On-Line"→ "Temperature Calibration."
   o   Enter the known temperature and click "Start."
   o   When the temperature, seen on the graph and numerically above the graph, has stabilized click "Stop."
   o   Press "Yes" to update the instrument.

(5) Repeat step 3 several times until the instrument-measured temperature agrees with the known input temperature.

(6) Click "Done."

Zero values or other anomalous acoustic-stage values in the data files downloaded from a Nortek EasyQ or OTT SLD may be the result of a poorly calibrated pressure transducer or a pressure transducer that has drifted out of calibration. To correct this problem, measure the distance from the upward looking acoustic transducer to the water surface (in meters). Enter this value in the "Set Pressure Offset" window: "On-Line"→ "Set Pressure Offset."

Steps for ADP troubleshooting, if the instrument is unresponsive:

(1) Check power source to ensure that it is providing 12 volts.

(2) If the instrument is connected via radio modems, connect it directly to the instrument cable.

(3) Check all communication settings. Settings should be for the proper serial port and baud rates of 38,400 for the Nortek EasyQ and 9,600 for the newer OTT SLDs.

(4) Try to connect with a different computer.

(5) Remove the instrument from the water and try to connect to it with a different communication cable.

(6) Send the instrument in for service if steps 1 through 5 do not reestablish communication.

Appendix 2: Two-Way Satellite-Telemetry System

Satellite-Telemetry System Instructions

Instructions for Connecting to, Checking, Downloading Data at, and Transferring Data from Remote Sites with the Satellite-Telemetry System

At the preset time when the main onsite computer (PC) turns on, double-click the pcAnywhere remote control icon corresponding to the station that you want to access.

(1) A window will pop up "pcAnywhere Waiting…"; this window will disappear when connection to the site is established. Once a connection is established, the desktop view of the "remote" onsite PC will be visible on your desktop, additional controls for managing the remote session will be visible on the left of the screen. **IMPORTANT:** using the remote desktop requires patience because of the lag times associated with the satellite connection.

(2) The program (SatControl) that controls when, and for how long, the onsite PC turns on will be visible.
   o On the onsite PC desktop there are shortcuts to various programs, HyperTerminal links to the different instruments (located at the top of the window, a shortcut to the torugsetpoints.txt file, and a com ports.txt file describing what instruments are on which com ports).

(3) Use the AddTime button in SatControl to add enough time, usually 40 minutes, to check the instruments, download data from the instruments, relaunch the instruments, and transfer the downloaded data from the onsite PC to in-office computers, using FTP. The current SatControl operation is shown in purple; it should say "AddTime," and then additional minutes will be displayed to the right of the AddTime button. If SatControl does not respond, wait several minutes and try again. If it still does not respond, close and re-open SatControl (shortcut located on remote desktop). SatControl will sometimes run a replace program resulting in the SatControl window temporarily disappearing. This is normal.
   o Displayed on SatControl will be the time remaining before the system shuts down, the current date and time, the battery voltage, the RTU temperature, the PC box temperature, the modem box temperature, and the ambient temperature. Along with the information displayed, there are three buttons: "AddTime," "Turn Off PC/Modem," and "Exit." Clicking the "AddTime" button will add time in 10-minute intervals to the amount of time remaining, until the satellite-telemetry system PC and modem

shut down. Clicking the "Turn Off PC/Modem" will shut down the system, clicking "Exit" will exit SatControl.
   o Expanding the SatControl panel will reveal more information and options (see Instructions for Using the SatControl Program below)

(4) Check the onsite PC time and update if needed; instrument clocks will be set from the onsite PC's time so it is important that this time is correct.

(5) Check for any Microsoft updates.
   o Updates appear as a yellow shield icon on the right of the "start" bar at the bottom of the screen.
   o Make sure that all updates are fully installed before the system is shut down.

(6) Check the LISST–100, if present. Because the vendor-supplied software for communicating with a LISST–100 does not fully function over the USB cables that are integral to the satellite-telemetry system, instructions for downloading data from and relaunching a LISST–100 over the satellite-telemetry system are provided in appendix 5: Sequoia Scientific LISST–100. If a LISST–100 is present at the sediment-transport monitoring station you are logged into, open up the corresponding HyperTerminal window (located at the top of the onsite PC desktop). **IMPORTANT:** failure to open the LISST–100 HyperTerminal windows may result in a Control-C command being inadvertently sent to the LISST–100, shutting it down, as other instruments try to query the com port occupied by the LISST–100. After the LISST–100 HyperTerminal icon is opened, a window with information on the time to next sample will appear. If nothing is displayed, wait several minutes to see if the instrument was sampling. If still nothing is displayed, check to make sure you are pointed to the proper com port and then consult appendix 5: Sequoia Scientific LISST–100.

(7) Check the RUG Trigger (30-mile and 61-mile sites only). Open the RUG Trigger HyperTerminal window and type (without quotes) "c,7" to display current status. Note the voltage and last turbidity value measured by the YSI sonde. Close the window by clicking the hang-up-phone icon (fourth icon from the left) at the top of the HyperTerminal window and closing the window. Other RUG Trigger commands are listed in appendix 3: RUG ISCO Trigger.

(8) Check the ISCO pump sampler(s). Open the ISCO(s) HyperTerminal window and type (without quotes): "??" CR (enter), "menu" CR, "st" CR, "q" CR, "q" CR. Record the sampler status, the bottle number, and whether the ISCO pump sampler is currently enabled or disabled. Typing the last q should get you out of all

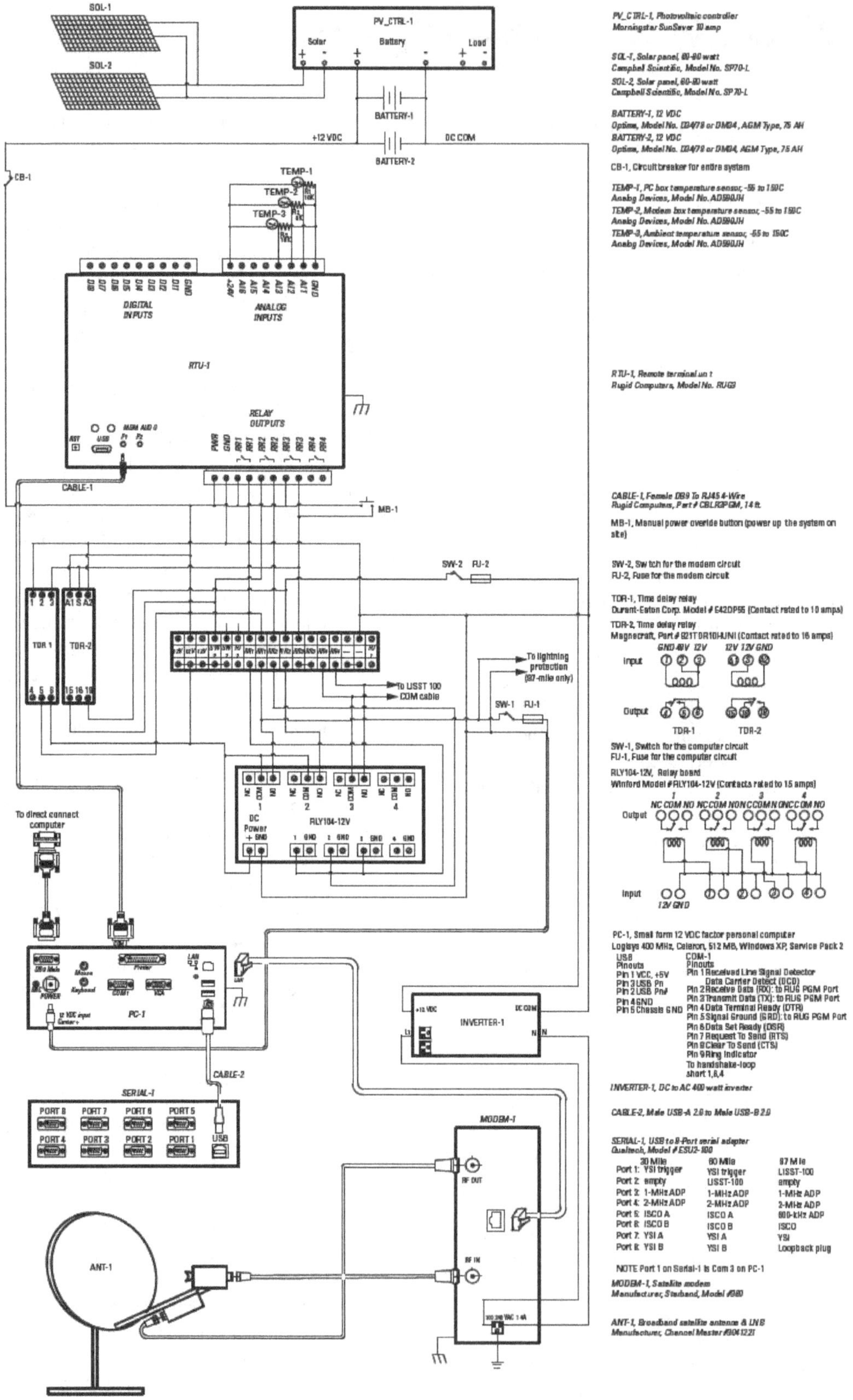

Figure 2–1. Wiring-connection diagram of the two-way satellite-telemetry system. See attached file for a large format version of this figure.

"carrot" prompts. Exit the HyperTerminal window as described in step 7.

o The first command "??" enters you into the ISCO pump-sampler control mode and displays information about the ISCO pump sampler as well as a "carrot" prompt.

o The menu command brings up a list of options: status, screen dump, pause, disable, enable, take sample, control, and quit. **IMPORTANT:** the use of commands other than those described herein may result in loss of sample information; use only the status (st) and quit (q) commands unless you have a thorough understanding of the ISCO pump sampler.

o The "st" (status) command will display the current status of the ISCO pump sampler.

o The "q" CR "q" CR commands will back the user out of the ISCO pump sampler menus, ending with no "carrot" prompt.

(9) Check the YSI sonde(s). Open the YSI sonde(s) HyperTerminal window and press CR (enter) until you see a # prompt. Type (without quotes): "menu" CR, "4", Esc (escape key), Esc, "Y". Record the YSI sonde status and voltage. After this, the # prompt should reappear. Exit the HyperTerminal window as described in step 7.

o The menu command brings up a list of options: Run (1), Calibrate (2), File (3), Status (4), System (5), Report (6), Sensor (7), and Advanced (8). **IMPORTANT:** the use of commands other that those described herein may result in loss of data; use only the "4" (status) and "0" (previous menu) commands unless you have a thorough understanding of the YSI sonde.

o The "4" (status) command will display the current status of the YSI sonde.

o The "0" or "Esc" (previous menu) commands will back the user out of the YSI-sonde menus, ending with the # prompt.

o A more detailed description and downloading instructions for the YSI sonde can be found in appendix 3: RUG ISCO Trigger.

(10) Wake up the MaxStream (Digi) radio modems connected to any remote ADPs (if present). Note that this step does not need to be completed when ADPs are directly attached to the satellite-telemetry system at the primary site.

o ADPs remotely connected via radio modem to the primary site (30-mile 1 MHz, 30-mile 2 MHz, and 61-mile 1 MHz) need to have the modems woken up before these instruments can be queried. To wake the modems up, open the HyperTerminal window corresponding to the ADP and hold down any key until either "Confirmed" or "§" is returned. After "Confirmed" or "§" is returned, any key

stroke will result in the return of a symbol similar to "§." Exit the HyperTerminal window as described in step 7; failure to close the HyperTerminal window will cause the corresponding com port to be occupied when attempting to download data from the ADPs, using the ADP-vendor-supplied software. The radio modems are in power-save mode and look for a data transmission every 16 seconds. The radio modems will stay awake for several minutes of nonuse and then return to power-save mode.

(11) Check the 1- and 2-MHz ADPs. Open the Nortek-supplied EasyQ program (located on the onsite PC desktop just under the HyperTerminal icons). Simultaneously press the (without quotes) "Alt" and "s" keys or pull down "Communication" and select "Serial Port." Select the com port for the instrument you are querying (1-MHz ADPs are usually connected to com 5, and 2-MHz ADPs are usually connected to com 6; check the list of com ports on the remote desktop if needed). Pull down "Deployment." Select "Recorder Data Retrieval." A pop up window will appear with the message "The instrument is collecting data. Do you want to stop data collection? y/n." If you are only checking to see if the instrument is working, press "n" and exit the program or select a different instrument.

(12) Check the 600-kHz ADP (87-mile only). Follow the instructions as for the ADP but use the Nortek-supplied AquaPro program.

(13) Transfer files downloaded from the instruments to the onsite PC to your office computer. Under "Session Manager" to the left of the onsite PC desktop, select "File Transfer." This will load a page with your office computer on the left and the onsite PC on the right. Select the files to transfer and drag them to the location you want to transfer them to (you can also use the file transfer arrows located between the two windows). A pop-up window will appear asking if you are sure—click yes. File transfer progress is shown at the bottom of the window. When all the files to be transferred have been added to the file transfer queue, select "Remote control" under the "Session Manager." The desktop view of the onsite PC will appear and then go black and re-appear; this is normal. Check to see if all the files to be transferred have appeared on your office computer.

(14) If all instruments have been checked and are running, and/or data have been downloaded and the instruments relaunched, shut down the remote session. Close any LISST–100 or other HyperTerminal windows. Click "Turn off PC/Modem" in SatControl. The button should change to "Exiting SatControl," and the time remaining will change to 3 minutes. Close the "pcAnywhere Remote" window.

Instructions for Using the SatControl Program

Control buttons found in the expanded SatControl window include:

AddTime—Adds 10 minutes to the time the system stays on

Turn Off PC/Modem—Shuts down the system.

Exit—Exits the SatControl program.

Read Labels From RUG—Reads the setpoints off the RUG and displays them in the SatControl program under Setpoints.

Write Labels to RUG—Sends the setpoints displayed on the right under the Setpoints heading to the RUG; users can change these setpoints and write them to the RUG. These changes will not change the setpoints file so the next time the PC boots the setpoints will revert back to those in the setpoints file.

Get Setpoints From RUG and Write to File—Reads the setpoints from the RUG and writes them to the setpoints text file. Clicking this button after making setpoint changes and writing them to the RUG will make the changes permanent. The time that the satellite-telemetry system turns on can also be permanently changed by changing the setpoints text file directly. A shortcut to the setpoints text (torugsetpoints.txt) file can be found on the desktop of the onsite PC. **IMPORTANT:** if a mistake is made and saved into the setpoints file (for example, a missing decimal point or space) the satellite-telemetry system will likely not turn on again, requiring a visit to the station.

Put Setpoints to RUG From File—Sends the setpoints from the text file to the RUG.

Maintain Power for RUG Programming—Keeps the system power on during remote RUG reprogramming events

Kill Modem—Turns off the satellite modem

Kill PC—Turns off the remote PC

Send RTC to RUG—Sends the onsite PC time to the RUG. This must be done periodically to keep the RUG time, and the time the system turns on, from drifting.

Get Entire Log—Recovers the internal log from the RUG.

Get Current Conditions—Gets the current conditions from the RUG.

Clear Log—Clears the internal RUG log.

Setpoints found in the torugsetpoints.txt file—a shortcut to this file can be found on the desktop; the setpoints file is located at C:\gcmrc\satsys\control\ or on the SatContol program (these setpoints will vary by station):

BattLo, 11.2
BattLoDef, 11.2
HRlogDef, 0.0
HRlogSP1, 0.0
HRmodemONDef1, 9.0
HRmodemONDef2, 9.0
HRmodemONSP1, 11.0
HRmodemONSP2, 11.0
HRpcONDef1, 9.0
HRpcONDef2, 9.0
HRpcONSP1, 11.0
HRpcONSP2, 11.0
killModemDelay, 180.0
killPCDelay, 180.0
LogFrequencyDef, 3600.0
MINmodemONDef1, 55.0
MINmodemONDef2, 55.0
MINmodemONSP1, 25.0
MINmodemONSP2, 25.0
MINpcONDef1, 55.0
MINpcONDef2, 55.0
MINpcONSP1, 28.0
MINpcONSP2, 28.0
ModemBoxHiTempDef, 170.0
ModemBoxHiTempSP1, 170.0
modemOFFdelayDef1, 900.0
modemOFFdelayDef2, 900.0
modemOFFdelaySP1, 1080.0
modemOFFdelaySP2, 1080.0
PCBoxHiTempDef, 170.0
PCBoxHiTempSP1, 170.0
pcOFFdelayDef1, 900.0
pcOFFdelayDef2, 900.0
pcOFFdelaySP1, 900.0
pcOFFdelaySP2, 900.0
RugHiTempDef, 170.0
RugHiTempSP1, 170.0
ExtOFFdelaySP, 600.0
ExternalTempOffset, 100.4
ModemTempOffset, 101.5
PCTempOffset, 101.0

Instructions for Onsite Satellite-Telemetry System Troubleshooting

(1) Check the RUG Time. The UP arrow may need to be pressed on the RUG for the RUG to display the time.

(2) Press the "_" button on the RUG.

(3) Press the "2" button on the RUG to get to Setpoints.

(4) Scroll down to HR MODEM ON SP1. Press "CLEAR."

(5) Type in a setpoint; (for example, 17.0 for 5 PM). Make sure to put in the decimal point and the 0. Press the enter button.

(6) Scroll down to HR PC ON SP1. Press "CLEAR."

(7) Type in a setpoint; (for example, 17.0 for 5 PM). Make sure to put in the decimal point and the 0. Press the enter button.

(8) Scroll down to MIN MODEMon SP1. Press "CLEAR."

(9) Type in a setpoint; (for example, 30.0 for half past the hour). Make sure to put in the decimal point and the 0. Press the enter button.

(10) Scroll down to MIN PC ON SP1. Press "CLEAR."

(11) Type in a setpoint; (for example, 30.0 for half past the hour). Make sure to put in the decimal point and the 0. Press the enter button.

(12) At the time entered, the RUG should fire the relay board. Three red LEDs on the relay board should light.

(13) If you know the modem is working, look and listen for the PC to boot up.

(14) If the PC does not boot up, pull the power cable out of the back of the PC and put the red voltmeter lead in the inside of the plug and the black lead on the outside of the plug. The voltmeter should read around 12 volts, or higher. If there is power to the PC, go to number 15. If the PC is not getting power, this could be a problem with the RUG output, a loose wire on the relay board, a bad relay on the relay board, a bad cable/connection on the PC, or a blown PC line fuse.

(15) If the PC does not boot up, but the PC has power, it could be that the shutdown sequence of SatControl did not work properly because of a Windows update running or a Windows operating system malfunction. Try turning "on" the PC with the power button located on the front of the PC. If the system appears to boot normally, try pcAnywhere. If pcAnywhere is able to direct connect to the PC, check the system time and date on the PC. If the time and date are far off, go to number 16. If the time and date are correct, try to shut down the system with SatControl. After the PC and modem go off, put in new setpoints on the RUG and try to bring up the system normally.

(16) If the system date and time are incorrect, reset the system time and date to the correct values. After setting these, restart the PC through the Windows start menu and see if the correct time is saved when the PC reboots. If the new time is not saved, the internal CMOS battery voltage is probably low. A new battery (model # CR2032) and an external monitor and keyboard will be needed to fix this.

(17) If the PC boots up, try the pcAnywhere direct connect to see if pcAnywhere can log into the PC. It will take several minutes for the PC to boot up normally, and possibly much longer if the PC was shut down during a Windows update. If the system date and time are far off, reset the system time and date to the correct values and go to number 16.

(18) If the PC tries to boot up, but pcAnywhere cannot connect with the PC, press the reset button on the PC, while it is powered up, to see if the PC will boot up properly. If this does not work, an external monitor will be required to check the status of the PC.

Setup of the Remote PC

Make sure that Windows on the PC is up to date; this will prevent lengthy downloads of updates when the PC is first deployed in the field.

Change BIOS so PC Boots After Power Failure

To change the PC so that it will power on automatically when the RUG applies power to the system, the Basic Input/Output System (BIOS) firmware on the PC needs to be changed. **IMPORTANT:** the PC will not boot up if the BIOS is corrupt; be very careful when making changes to the BIOS. Note that when you first turn on the PC, the BIOS manufacturers name will appear (normally AMI, Award, or Phoenix). Write this name down for future reference. Most PCs have slightly different ways to change the BIOS; this is how the USGS GCMRC PCs are configured. Other known possibilities are provided here.

(1) When the PC is starting up, press and hold the "Del" key on the keyboard.
  o Other PCs may use F1 or F2; if you do not know, watch the startup screen, as the command to enter the BIOS is usually displayed.

(2) This will open the BIOS; use the arrow keys to navigate, and press "enter" to select an item to change.

(3) Find "Integrated Peripherals."

(4) Find "PWRON After PWR-Fail", and press "enter" to make a change.
  o There are three options: off, on, and former status.

(5) Select "on" and press "enter."

(6) Press the "Esc" key to move back to the main menu.

(7) Save and Exit Setup (F10).
  o If there is no save option, try pressing the "Esc" key.

Change the Interval that Windows Updates the PC Internal Time

Windows updates the time on the PC once a week by default. The PC time may drift several minutes during this period; if this is the case, shorten the period between time updates. NOTE: these instructions are for Windows XP.

(1) Backup Windows registry; changing the registry may damage your operating system so be sure to back it up.
  o Click "Start" and select "Run."
  o Type "%SystemRoot%\system32\restore\rstrui.exe", and click "ok."
  o Or Select Start button→ All Programs→ Accessories→ System Tools→ System Restore.
  o Select "Create a Restore Point" and click "next."
  o Enter a descriptive name and click "Create."
  o Close.

(2) Start button→ run, or press the "Windows key" plus the "r" key.

(3) Type "regedit" and press "enter."

(4) Navigate to "HKEY_LOCAL_MACHINE\SYSTEM\ControlSet001\Services\W32Time \TimeProviders\NtpClient."

(5) Select "SpecialPollInterval."

(6) Change decimal value (measured in seconds) from 604800 (7 days) to 172800 (2 days) or 86400 (1 day).

(7) Click "ok" and exit regedit.

Load SatControl Program

(1) Check PC for Microsoft .NET framework, if it is not present load the program.
  - o .NET framework can be found online at http://www.microsoft.com/NET/.

(2) Copy SatControl, related programs, and related text files into C:\gcmrc\satsys\control.

(3) Manually install gcmrcService.
  - o Open the command window.
    - ▪ Start button→ run→ type "cmd."
    - ▪ Command window will open.
  - o Type "sc create satcontrol binpath= C:\gcmrc\satsys\control\gcmrcService.exe" then press "enter."
    - ▪ Pressing "enter" should return with [SC] CreateService SUCCESS.
    - ▪ To delete a service type "sc delete ServiceName" followed by "enter."
  - o Close command window.

(4) Change SatControl service settings to load the program on startup and have it be visible on the desktop.
  - o Start button→ Control Panel→ Administrative Tools (if not visible switch to classic view)→ Services → double click on "satcontrol."
  - o Under the "General" tab, change startup type to "Automatic."
  - o Under the "Log On" tab, check the box next to "Allow service to interact with desktop."
  - o Press the "OK" button and close the Services window.

(5) Reboot the PC; SatControl should appear on the desktop within several minutes.
  - o If the PC is not connected to a RUG, all the fields will be blank.

Other Setup Steps

(1) Check IP address assignment
  - o Start button→ Control Panel→ Network Connections→ right click on "Local Area Connection."
  - o Select "properties"; this will open a window.
  - o Highlight "Internet Protocol (TCP/IP)."
  - o Press "properties" button; this will open a second window.
  - o Check "Obtain an IP address automatically", press "OK."
  - o Close windows.

(2) Set LAN settings

  - o Start button→ Settings → Control Panel→ Internet Options.
  - o Under the "Connections" tab click the "LAN Settings" button.
  - o A window will appear with "Local Area Network (LAN) Settings"; make sure that the "Automatic configuration" and "Proxy server" boxes are unchecked.

(3) Load and set up pcAnywhere.
  - o Load the pcAnywhere software; make sure that you install the host version on the remote PC.
  - o On the host machine (found in the pcAnywhere Manager) under hosts, right click on "Network, Cable, DSL" and select "Properties."
  - o Under the "Connection Info" tab check the COM2 and TCP/IP boxes.
  - o Under the "Settings" tab "Host Startup," check the "Launch with Widows" and "Run Minimized" boxes.
  - o Under the "Callers" tab, right click on the callers list and press New→Item…
    - ▪ Enter a user name and password; this name will appear on the "Callers list."
  - o Under the "Security Options" tab and "Login options," change the maximum allowed login attempts per call to 50.

(4) Load all instrument manufacture programs and set to the appropriate com port.

Radio Modem Setup and Operation

These instructions are for MaxStream (Digi) XTend PKG radio modems.

(1) Set the dip switches.
  - o On the end of the modem with the antenna port, set the dip switches 1, 5, and 6 in the "up" position. This sets the modem for RS–232 input with user-defined receive and transmit modes.

(2) Attach an antenna (even if the modems are being tested right next to each other, they might not work without antennas), connect the modem to a 12-volt power source, and connect the modem to a PC loaded with the vendor-supplied X-CTU program using a serial cable. If testing the modems be aware that if the modem power level is set to 1 watt, and the modems are operated in close proximity, the units may be damaged.

(3) Program the modem using the X–CTU program. The following settings have been changed in the "Modem Configuration," all other settings have been left at factory settings (table 2–1). For a full description of all settings refer to the XTend Product Manual. Click on the "Read" button to load the current modem settings.

  - o ID—Modem VID; set each pair of base and receiving modems to the same ID (default ID is 3332).

o   HP—Hopping Channel; set each pair to a hopping channel (0–9).

o   DT—Destination Address; set each pair to a destination address (0–FFFF). If set to FFFF, the base and remote settings are the same.

o   MY—Source Address; set each modem with its own source address. For simplicity make each modem in a pair have a source address that is one off from the other; make the base the lower number (0–FFFF).

o   MK—Address Mask; set the same as DT.

o   RR—Retries, defaults to 0. Set to 8 or more if experiencing data transmission problems that are not fixed by increasing the power level. To use RR, the MK and DT values must not be set to FFFF.

o   MT—Multiple-Transmit; set to 3 if experiencing data transmission problems and are using MT=FFFF.

o   RN—Delay Slots; set to 4 on the remote modem if there are possible interference issues; RR must be enabled.

o   BD—Baud Rate; set to the baud rate of the ADP, 5 (38,400) for Nortek ADPs or 3 (9600) for OTT ADPs.

o   PL—TX Power Level; to conserve power, set to the lowest level possible with good transmission.

o   SM —Sleep Mode; set the base to 2 – Serial Port Sleep and the remote to 8 – Cyclic 16 seconds (the modem "wakes" every 16 seconds to look for transmissions).

o   ST—Time before Sleep; set to BB8 (approximately 5 minutes of inactivity before the modem goes to sleep).

(4)   Load Program onto modem. If the programming hangs, the reset button on the modem may need to be pressed (small black button located to the left of the serial cable IO).

(5)   Test that the set of modems is communicating and perform a range test.

o   Attach a loop-back plug to the serial port of the remote modem.

o   In the X-CTU software under the "Range Test" tab, click "Start."

o   An output of "0123456789:;<=>?@ ABCDEFGHIJKLMNO" will stream down the window.

o   Check the RSSI window for a graphical display of the signal strength; try to maximize this by moving antennas and, if necessary, changing modem settings.

(6)   Once the modem setup is functioning properly, attach the remote modem to an ADP (must have a null modem adaptor between the ADP and the modem) and the base modem to the satellite-telemetry system. Check to ensure that the ADP is functioning properly through the modem (some older model Nortek ADPs do not function with these radio modems and may need to be modified by Nortek).

Lightning Surge Protection

At the 87-mile station, a USB extender is used because the serial adapter is located beyond the 5-m range of USB cables. To prevent the USB extender from burning out, the power pin of the USB cable (Pin 1) is removed.

Table 2–1. Radio-modem settings for all stations.

[Using different modem firmware versions will change the settings available ; -, denotes parameters that can not be changed; %, percent]

Programming parameters	Radio modem location (sediment-transport monitoring station) and instrument					
	30-mile 1-MHz base	30-mile 1-MHz remote	30-mile 2-MHz base	30-mile 2-MHz remote	61-mile 1-MHz base	61-mile 1-MHz remote
ID – Modem VID	3332	3332	3333	3333	3332	3332
HP – Hopping channel	0	0	0	0	0	0
DT – Destination address	FFFF	FFFF	0	0	0	0
MY – Source address	5	6	7	8	FFFF	FFFF
MK – Address mask	FFFF	FFFF	0	0	FFFF	FFFF
RR – Retries	8	8	8	8	A	A
MT – Multiple transmit	3	3	3	3	0	0
RN – Delay slots	0	4	0	4	0	0
TT – Streaming limit	0	0	0	0	0	0
KY – AES encryption key	0	0	0	0	0	0
BD – Baud rate	5	5	5	5	5	5
NB – Parity	0	0	0	0	0	0
SB – Stop bits	0	0	0	0	0	0
RB – Packetization threshold	800	800	800	800	800	800
RO – Packetization timeout	3	3	3	3	3	3
PK – Maximum RF packet size	800	800	800	800	800	800
CS – Pin 9 configuration	0	0	0	0	0	0
TR – Pin 10 configuration	0	0	0	0	0	0
CD – Pin 3 configuration	2	2	2	2	2	2
FL – Software flow control	0	0	0	0	0	0
FT – Flow control threshold	BEF	BEF	BEF	BEF	BEF	BEF
BR – RF data rate	1	1	1	1	1	1
PL – TX power level	3	3	3	3	3	3
TX – Transmit only	0	0	0	0	0	0
FS – Forced sync time	0	0	0	0	0	0
VR – Firmware version	-	-	-	-	-	-
HV – Hardware version	-	-	-	-	-	-
SH – Serial number high	-	-	-	-	-	-
SL – Serial number low	-	-	-	-	-	-
RP – RSSI TWN timer	20	20	20	20	20	20
TP – Board temperature	-	-	-	-	-	-
%V – Board voltage	-	-	-	-	-	-
DB – Received signal strength	-	-	-	-	-	-
ER – Receive error count	-	-	-	-	-	-
GD – Receive good count	-	-	-	-	-	-
TR – Delivery failure count	-	-	-	-	-	-
SM – Sleep mode	2	8	2	8	1	8
ST – Time before sleep	E10	BB8	E10	BB8	E10	BB8
HT – Time before wake-up initializer	BB8	FFFF	BB8	FFFF	BA4	FFFF
LH – Wake-up initializer timer	AA	1	AA	1	AA	1
PW – Pin wake-up	0	0	0	0	0	0
BT – Baud time before	A	A	A	A	A	A
CC – Command sequence character	2B	2B	2B	2B	2B	2B
AT – Guard time after	A	A	A	A	A	A
CT – Command mode timeout	C8	C8	C8	C8	C8	C8

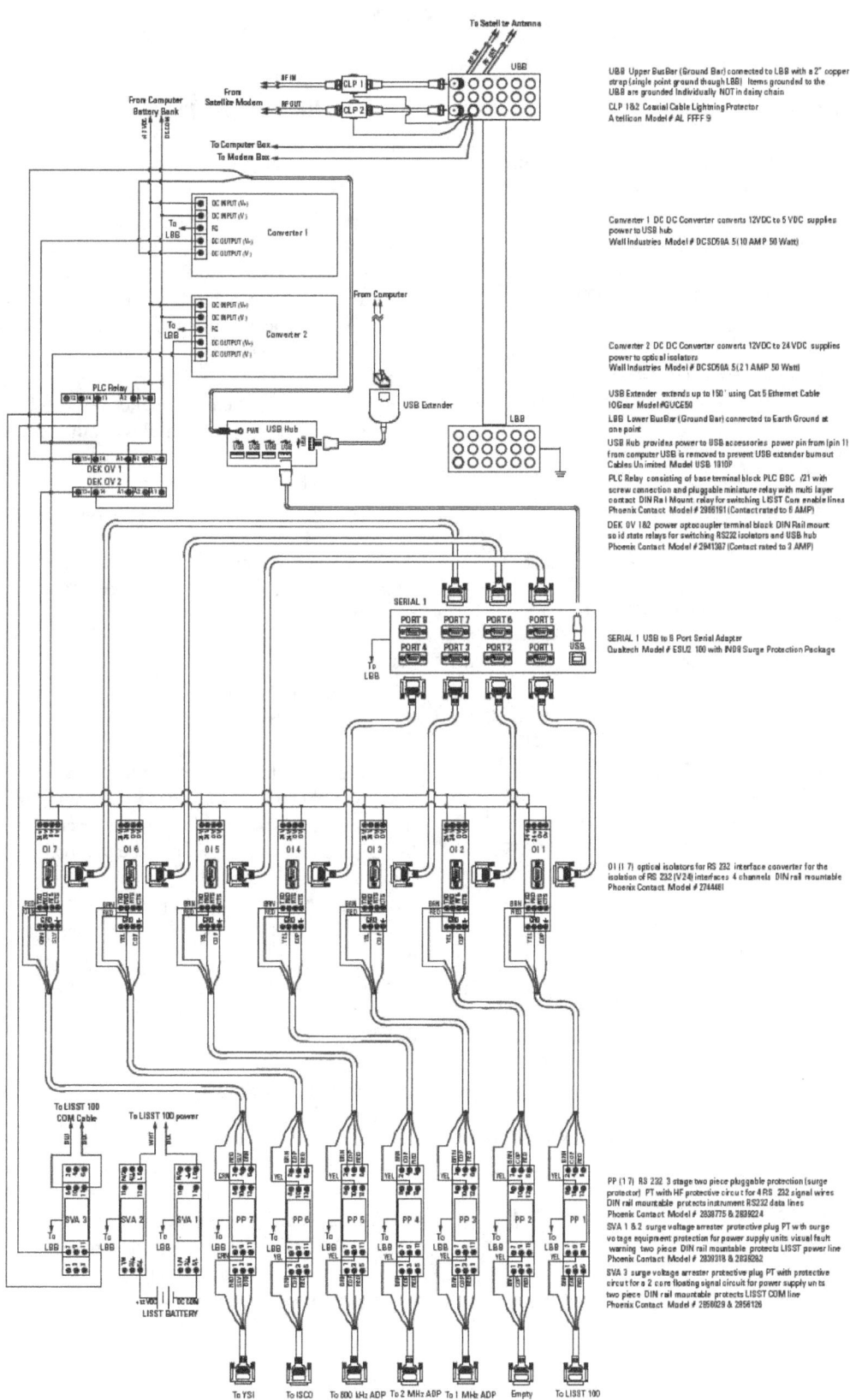

Figure 2–2. Wiring-connection diagram of the lightning-protection system. See attached file for a large format version of this figure.

Appendix 3: RUG ISCO Trigger and ISCO 6712 Automatic Pump Sampler

RUG ISCO Trigger Commands

The commands for the RUG ISCO Trigger are entered though the RUG HyperTerminal window. The commands are:

- c,1
  - o Displays all commands.
- c,2
  - o Displays all setpoint commands.
- c,3
  - o Set internal RUG clock. Set clock as: YYYY,MMDD,HHMM,SS,DW; DW is the day of the week starting on Sunday = 1.
- c,4
  - o Displays setpoints.
- c,5
  - o Displays default setpoints.
- c,6
  - o Set setpoint defaults.
- c,7
  - o Displays current system status.
- c,8
  - o Force the YSI sonde to take a measurement.
- c,9
  - o Display the log report. This will scroll the entire log across the screen. To capture the log, select from the HyperTerminal window→"Transfer"→"Capture Text" before displaying the log. After the entire log has scrolled, select from the HyperTerminal window→"Transfer"→"Capture Text"→"Stop."
- c,10
  - o Clear the log report file.
- c,11
  - o Run program. Starts the RUG program.
- c,12
  - o Suspend program. Suspends the RUG program.
- c,13
  - o Enable the ISCO. Starts the ISCO sampling program.
- c,14
  - o Disable the ISCO. Suspends the ISCO sampling program.
- c,15
  - o Cycle RUG display on.
- c,16
  - o Cycle RUG display off.

Troubleshooting Instructions for the ISCO Pump Sampler

Most common problems with the ISCO 6712 automatic pump sampler relate to the liquid detector not functioning properly. This is typically the result of the pump tube not being seated correctly on the liquid detector, an old pump tube, or calibrating the sample volume too frequently. If a "no liquid detected" error occurs, the pump tube should be reseated. If that fails to fix the problem, the pump tube should be replaced. (Note: storing pump tubes in very hot environments will reduce their lifespan.) In hot environments, pump tubes should be replaced every 6 months regardless of the degree of usage. If the pump error remains after the above steps, the pump table may have to be reset. To reset the pump table, change the pump tube line length (note this number before you change it), and then change the pump tube line length back to the original length. During this process, you should see "generating pump tables"; after you complete this process, you will need to recalibrate the sample volume.

The ISCO pump sampler has an internal battery for the clock and for storing the ISCO program. This internal battery needs to be replaced every several years. The ISCO pump head can be sent in for replacement of this battery or, if necessary, the battery may be replaced in the field (contact Teledyne/ISCO for replacement batteries). **IMPORTANT:** The internal battery is soldered in a location that is hard to reach and requires removing several circuit boards from the ISCO pump sampler. Only replace the internal battery yourself if you are familiar with the internal workings of the ISCO and are confident in your soldering skills.

In the event that the ISCO pump sampler "locks up" or needs to be reset, a "hard reboot" can be executed. This will erase the ISCO program; write down the program settings before a hard reboot. The "hard reboot" instructions are as follows:

(1) Disconnect power.
(2) Wait 60 seconds.
(3) Simultaneously press and hold the stop button (red button with a triangle within a circle symbol) and enter button (yellow button with a leftward pointing arrow).
(4) While holding the buttons, reconnect the power and then release the buttons.
(5) Reprogram the ISCO.

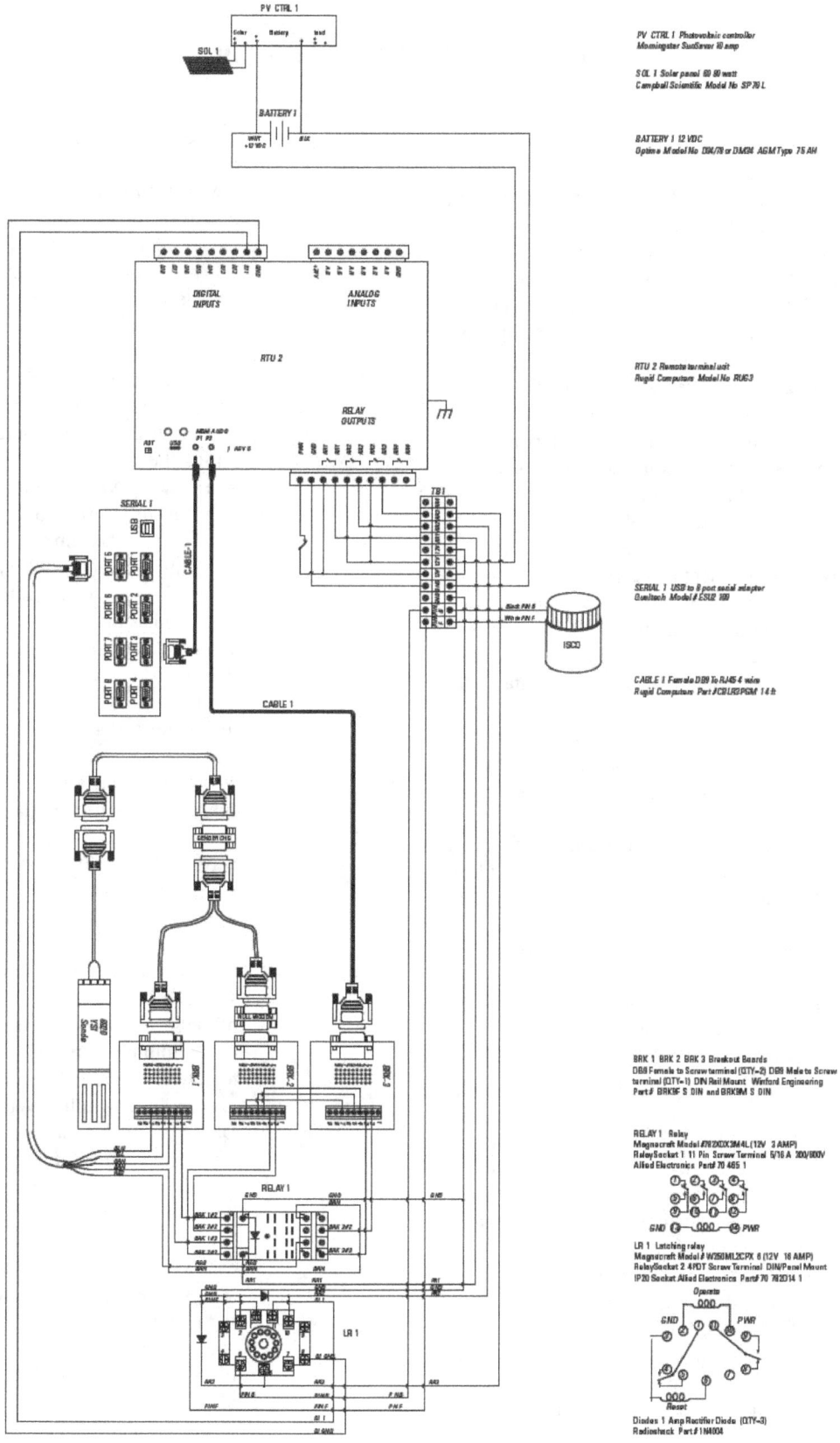

Figure 3–1. Wiring-connection diagram of the ISCO pump sampler RUG Trigger. See attached file for a large format version of this figure.

Appendix 4: Example of a Typical River-Trip Work Plan

Work plan for 30-mile station—first day.

- Travel to site.
- Set up temporary tagline for EWI measurements and ADCP discharge measurements. Measure the width of the river and calculate five equal-width increments (even though the USGS-recommended minimum number of sampling verticals is 10, comparisons of EWI measurements collected from the Colorado River in Marble and Grand Canyons with 5 sampling verticals to those collected with 10 sampling verticals show that they are equivalent). Mark the center of each increment with a glow stick (allowing for collection of suspended-sediment samples at night).
- Clean ADP transducers, inspect ADP mounts, and check remote ADP battery banks.
- Record a tape-down measurement at each ADP. The tape-down measurement is recorded as the distance from the water surface to a reference point measured with an engineer's folding ruler.
- Download data from ADPs over the satellite-telemetry system accessed on site. Check the ADP data for consistency.
- Perform routine maintenance at the station, including emptying the ISCO pump sampler, checking the battery voltage on all battery banks, and inspecting the station.
- Clean or replace the ISCO pump-sampler intake tubes. Cleaning the tubes consists of pumping bleach from the ISCO down through the intake tube to remove any biological growth.
- Check, and replace if necessary, the ISCO peristaltic pump tube.
- Restock the station with any supplies, including ISCO bottles, needed for routine hike-in maintenance.

Work plan for 30-mile station—second day.

- Make four EWI suspended-sediment measurements. Measurements are usually made every 3–4 hours beginning around 7 a.m. and concluding after 8 p m. Samples are collected over the entire day to ensure that a wide range of discharges and suspended-sediment concentrations are represented.
  - Suspended-sediment measurements are collected using a US D–96–A1 depth-integrating sampler equipped with a ¼-inch (0.635 cm) nozzle. US D–96 samplers may be used during higher-than-normal dam releases. These samplers are deployed by using B-reels and an aluminum crane mounted on the front of a motorized boat (fig. 11).

- At the five EWI sampling verticals, the depth of the river is sounded prior to each EWI measurement.
- Electronic metronomes are used to ensure the steady transit required for EWI samples. The pace of the metronome is adjusted to increase or decrease sample volumes, as needed, within USGS-accepted transit rates (Edwards and Glysson, 1999).
- To prevent the sample from being contaminated with bed sediment, care is taken to prevent the sampler nozzle from hitting the riverbed.
- At each EWI vertical, the sampler completes four transits in order to increase time averaging of the sample and minimize error (Topping and others, 2011). As documented in Topping and others (2011), one transit is defined as the path a depth-integrating suspended-sediment sampler takes either from the water surface to the bed or from the bed to the water surface. Therefore, standard deployment of a depth-integrating sampler at a vertical, where the nozzle is open as the sampler is lowered to the bed and subsequently raised to the surface, consists of two transits. The sampler is emptied between transits two and three to prevent overfill of the sampler.
- The suspended-sediment samples collected at each sampling vertical in an individual EWI measurement are combined in plastic 4-liter (L) bottles. The cumulative volume of the sample collected is measured between EWI sampling verticals with a calibrated rod to determine the volume of sample collected at each EWI sampling vertical.
- Data from each measurement are recorded, including station name, measurement collection date, measurement collection personnel, boat operator, start and end times (to the nearest minute), depth of water at each vertical (to the nearest tenth of a foot), sampler transit times at each EWI sampling vertical (to the nearest tenth of a second), and sample volume collected from each EWI sampling vertical (to the nearest tenth of a liter).
- Upon completion of each EWI measurement, sample-container lids are taped closed to prevent sample spillage or contamination, and containers are labeled

with the station name, date and time of measurement collection, water level in the sample container, and the number of bottles that constitute the EWI measurement.

  o Collection of accurate EWI measurements from a boat requires that the boat operator be highly skilled and extremely conscious of boat position and velocity.

- Collect four ISCO pump samples from each pump sampler concurrently with the four EWI measurements.

  o During each EWI measurement, an ISCO pump sample is collected from each of the two pump samplers

  o To reduce the chance of bias, the order of the pumps used is switched between samples (the pump cannot be run simultaneously because of potential interference at the pump intakes).

- Make 3–4 ADCP discharge measurements. Discharge measurements are usually made every 3–4 hours beginning around 7 a m. and concluding after 8 p.m.

  o See section "Discharge Computations" for a discussion of the ADCP methods used.

- Collect one set of bed-material samples.

  o Bed-material samples are collected at each of the three center EWI sampling verticals. Bedrock or boulders underlie the outermost two EWI sampling verticals.

  o Bed-material samples are collected by using a pipe dredge (Edwards and Glysson, 1999) deployed from a motorized boat and dragged under the EWI sampling verticals. A pipe dredge is used instead of a US BM–54 sampler because of the coarse gravel that underlies some of the sampling verticals.

  o Bed-material samples are placed in cloth sample bags labeled with the station name, sample collection date, sample collection time, and EWI sampling vertical location.

- Calibrate YSI turbidity probes.
- Complete any necessary major repair or modification to the station.

Work plan for 30-mile station—final day.

- Make four EWI suspended-sediment measurements. Measurements are usually made every 3–4 hours, beginning around 7 a.m. and concluding after 5 p.m.
- Collect four ISCO pump samples from each pump sampler concurrently with the four EWI suspended-sediment samples.
- Make 3–4 ADCP discharge measurements. Discharge measurements are usually made every 3–4 hours, beginning around 7 a m. and concluding after 4 p.m.

- Collect one set of bed-material samples.
- Remove temporary tagline.
- Download and check all instruments.
- Ensure that the station is secure and all instruments are functioning normally.
- Depart site.

Work plan for 61-mile station—first day. All procedures are the same as those for the 30-mile station, except where noted.

- Travel to site.
- Set up temporary tagline for EWI measurements and ADCP discharge measurements.
- Clean ADP transducers, inspect ADP mounts, and check remote ADP battery banks.
- Clean LISST-100 with bleach and ultrasonic bath.
- Record a tape-down measurement at each ADP.
- Download data from ADP instruments over the satellite-telemetry system accessed on site; check the ADP data for consistency.
- Perform routine maintenance at station, including emptying the ISCO pump sampler, checking battery voltage on all battery banks, and inspecting the station.
- Clean or replace the ISCO pump-sampler intake tubes.
- Check, and replace if necessary, the ISCO peristaltic pump tube.
- Restock the station with supplies.

Work plan for 61-mile station—second day.

- Make four EWI suspended-sediment measurements. Measurements are usually made every 3–4 hours, beginning around 7 a m. and concluding after 8 p.m.
- Collect four ISCO pump samples from each pump sampler concurrently with the four EWI measurements.
- Make 3–4 ADCP discharge measurements. Discharge measurements are usually made every 3–4 hours, beginning around 7 a m. and concluding after 8 p.m.
- Collect one set of bed-material samples.

  o Bed-material samples are collected at each of the five EWI stations.

- Calibrate YSI turbidity probes.
- Complete any necessary major repair or modification to the station.

Work plan for 61-mile station—final day.

- Make four EWI suspended-sediment measurements. Measurements are usually made every 3–4 hours, beginning around 7 a m. and concluding after 5 p.m.
- Collect four ISCO pump samples from each pump sampler concurrently with the four EWI measurements.

- Make 3–4 ADCP discharge measurements. Discharge measurements are usually made every 3–4 hours, beginning around 7 a.m. and concluding after 4 p m.
- Collect one set of bed-material samples.
- Remove temporary tagline.
- Download and check all instruments.
- Ensure that the station is secure and all instruments are functioning normally.
- Depart site.

Work plan for 87-mile station—first day. All procedures are the same as those for the 30-mile station, except where noted.

- Travel to site, arriving midday.
- Clean ADP transducers, inspect ADP mounts, and check remote ADP battery banks.
- Clean LISST-100 with bleach and ultrasonic bath.
- Download data from ADP instruments over the satellite-telemetry system accessed on site; check the ADP data for consistency.
- Perform routine maintenance at station, including emptying the ISCO pump sampler, checking battery voltage on all battery banks, and inspecting the station.
- Clean or replace the ISCO pump sampler intake tubes.
- Check, and replace if necessary, the ISCO peristaltic pump tube.
- Make one EDI suspended-sediment measurement from the USGS measurement cableway.
    - Suspended-sediment measurements are collected using a US D–96–A1 depth-integrating sampler, equipped with a ¼-inch (0.635 cm) nozzle. US D–96 samplers may be used during higher-than-normal dam releases.
    - At the five EDI sampling verticals, the depth of the river is sounded prior to each EWI measurement.
    - To prevent the sample from being contaminated with bed sediment, care is taken to prevent the sampler nozzle from hitting the riverbed.
    - At each EDI sampling vertical, field personnel complete four transits with the sampler in order to increase time averaging of the sample and minimize error (Topping and others, 2011). As documented in Topping and others (2011), one transit is defined as the path a depth-integrating suspended-sediment sampler takes either from the water surface to the bed or from the bed to the water surface. Therefore, standard deployment of a depth-integrating sampler at a vertical, in which the nozzle is open as the sampler is lowered to the bed and subsequently raised to the surface, consists

of two transits. The sampler is emptied between transits two and three to prevent overfill of the sampler.
    - The suspended-sediment samples collected at each sampling vertical in an individual EDI measurement are combined in plastic 4-L bottles. The cumulative volume of the sample collected is measured between EDI sampling verticals with a calibrated rod to determine the volume of sample collected at each EDI sampling vertical.
    - Data from each measurement are recorded on a data sheet including station name, measurement collection date, measurement collection personnel, boat operator, start and end times (to the nearest minute), depth of water at each vertical (to the nearest tenth of a foot), sampler transit times at each EDI sampling vertical (to the nearest tenth of a second), and sample volume collected from each EDI sampling vertical (to the nearest tenth of a liter).
    - Upon completion of each EDI measurement, sample-container lids are taped closed to prevent sample spillage or contamination and containers are labeled with the station name, date and time of measurement collection, water level in the sample container, and number of bottles that constitute the EDI measurement.
- Collect one ISCO pump sample concurrently with EDI measurement.
- Collect a set of bed-material samples.
    - Bed-material samples are collected at each of the three center EDI sampling verticals.
    - Bed-material samples are collected from the USGS cableway by using a US BM–54 sampler.
    - Bed-material samples are placed in cloth sample bags labeled with the station name, sample collection date, sample collection time, and EDI sampling-vertical location.
- Calibrate YSI turbidity probes.
- Restock the station with supplies.

Work plan for 87-mile station—final day.

- Make two EDI suspended-sediment measurements.
- Collect two ISCO pump samples concurrently with the two EDI measurements.
- Collect one set of bed-material samples.
- Complete any necessary major repair or modification to the station.
- Download and check all instruments.
- Ensure that the station is secure and all instruments are functioning normally.

- Depart site next morning.

Work plan for 166-mile station—first day. All procedures are the same as those for the 61-mile station except for the absence of ISCO pump samplers at the 166-mile station.

- Travel to site, arriving midday.
- Set up temporary tagline for EWI measurements and ADCP discharge measurements.
- Clean ADP transducers, inspect ADP mounts, and check remote ADP battery banks.
- Record a tape-down measurement at each ADP.
- Download data from ADP instruments; check the ADP data for consistency.
- Perform routine maintenance at station, including checking battery voltage on all battery banks, and inspecting the station.
- Make two EWI suspended-sediment measurements 3–4 hours apart.

Work plan for 166-mile station—second day:

- Make four EWI suspended-sediment measurements. Measurements are usually made every 3–4 hours, beginning around 7 a m. and concluding after 8 p.m.
- Make 3–4 ADCP discharge measurements. Discharge measurements are usually made every 3–4 hours, beginning around 7 a m. and concluding after 8 p.m.
- Collect one set of bed-material samples.
- Complete any necessary major repair or modification to the station.

Work plan for 166-mile station—final day:

- Make four EWI suspended-sediment measurements. Measurements are usually made every 3–4 hours, beginning around 7 a m. and concluding after 5 p.m.
- Make 3–4 ADCP discharge measurements. Discharge measurements are usually made every 3–4 hours, beginning around 7 a m. and concluding after 4 p.m.

- Collect one set of bed-material samples.
- Remove temporary tagline.
- Download and check all instruments.
- Ensure that the station is secure and all instruments are functioning normally.
- Depart site next morning.

Work plan for 225-mile station—first day. All procedures are the same as those for the 87-mile station.

- Travel to site.
- Clean ADP transducers, inspect ADP mounts, and check remote ADP battery banks.
- Download data from ADP instruments; check the ADP data for consistency.
- Perform routine maintenance at station, including emptying the ISCO pump sampler, checking battery voltage on all battery banks, and inspecting the station.
- Clean or replace the ISCO pump-sampler intake tubes.
- Check, and replace if necessary, the ISCO peristaltic pump tube.
- Make one EDI suspended-sediment measurement.
- Collect one ISCO pump sample from each pump sampler concurrently with the EDI measurement.
- Collect one set of bed-material samples.
- Restock the station with supplies.

Work plan for 225-mile station—final day.

- Make one EDI suspended-sediment measurement.
- Collect one ISCO pump sample from each pump sampler concurrently with the EDI measurement.
- Complete any necessary major repair or modification to the station.
- Download data and check all instruments.
- Ensure that the station is secure and all instruments are functioning normally.
- Depart site.

Appendix 5: Instructions for Downloading Data from the Sequoia Scientific LISST–100 When Connecting Through the Satellite-Telemetry System

The LISST–100 offload Data Command Procedure is described below (C. Pottsmith, Sequoia Scientific, written commun., 2004). The following procedure assumes that the instrument is running the Base program provided with Version 4.3 or higher of the LISST–SOP Windows software. Communication and XMODEM transfer is at 9,600 baud, 8 data bits, no parity, and one stop bit. XMODEM is the 128-byte version. To offload data, open the LISST–100 HyperTerminal window.

(1) The running program must be stopped and the "OK>" prompt must be displayed. Pressing "enter" (CR) should return another "OK>" prompt. Pressing the "Ctrl" plus the "c" keys simultaneously (CTRL–C) should stop the program.

(2) Send "GOTO 15000" followed by a CR.

(3) LISST–100 returns the number of bytes to offload as ASCII characters.

(4) LISST–100 returns the text "WAITING FOR NAK."

(5) Press the "Ctrl" plus the "u" keys (CTRL–U) (15 Hex) within 60 seconds to start XMODEM transfer.

(6) Start Xmodem receive.

Once the data have been downloaded and checked, the Sequoia-Scientific-supplied LISST–SOP software can be used to erase the instrument memory, set the instrument clock, modify the deployment parameters, and launch the instrument.

www.ingramcontent.com/pod-product-compliance
Lightning Source LLC
Chambersburg PA
CBHW081622170526
45166CB00009B/3068